Lecture Notes in Computer Science 573
Edited by G. Goos and J. Hartmanis

G. Cohen S. Litsyn
A. Lobstein G. Zémor (Eds.)

Algebraic Coding

First French-Soviet Workshop
Paris, July 22-24, 1991
Proceedings

Springer-Verlag
Berlin Heidelberg New York
London Paris Tokyo
Hong Kong Barcelona
Budapest

Series Editors

Gerhard Goos
Universität Karlsruhe
Postfach 69 80
Vincenz-Priessnitz-Straße 1
W-7500 Karlsruhe, FRG

Juris Hartmanis
Department of Computer Science
Cornell University
5148 Upson Hall
Ithaca, NY 14853, USA

Volume Editors

Gérard Cohen
Antoine Lobstein
Gilles Zémor
École Nationale Supérieure des Télécommunications
46 rue Barrault, 75634 Paris Cedex 13, France

Simon Litsyn
Department of Electrical Engineering, Tel Aviv University
Ramat Aviv 69978, Israel

CR Subject Classification (1991): E.4

ISBN 3-540-55130-1 Springer-Verlag Berlin Heidelberg New York
ISBN 0-387-55130-1 Springer-Verlag New York Berlin Heidelberg

Printed in Germany

Typesetting: Camera ready by author
Printing and binding: Druckhaus Beltz, Hemsbach/Bergstr.
45/3140-543210 - Printed on acid-free paper

Preface

The idea of this first French–Soviet workshop on algebraic coding was born in Leningrad (now St. Petersburg) during the second international workshop on algebraic and combinatorial coding theory, in September 1990. Surprisingly, it actually took place some months later at the Ecole Nationale Supérieure des Télécommunications in Paris, July 22–24, 1991. Although scientists from Finland, Germany, Israel, Italy, Spain, and the United States also attended, the initial idea of having some of the best Soviet coding theorists was fully realized.

The papers in this volume fall rather naturally into four categories.

Applications of Exponential Sums

A. Tietäväinen shows a simple unified way of interpreting the autocorrelation of sequences as character sums and evaluating their minimum value. It can be applied to Kumar–Moreno and Lin and Komo sequences.

I. Shparlinski obtains new bounds for Gaussian sums

$$G_n(n,p) = \sum_{x \in GF(p)} \exp(2iax^n/p)$$

over prime finite fields that improve the classical bound $|G_n(n,p)| \leq np^{1/2}$ for certain values of n and p. Upper bounds are also provided for Gaussian sums over non-prime fields of the kind $GF(2^m)$. Finally, a natural extension of Gaussian sums over elliptic curves is considered, and upper bounds are obtained that shed light on the distribution of the generating points of the curve when its group is cyclic.

A. Barg presents new families of dc-constrained error-correcting codes. They have good minimum distance and running digital sum properties, which are proved through classical inequalities for character sums.

Covering Radius

A. Davydov describes new constructions for q-ary linear codes, represented by their parity-check matrices, providing infinite families of codes with covering radius 2 (for $q \geq 4$) and a table of the least known length of quaternary codes $[n, n-r]$ with covering radius 2 (for $r \leq 20$).

G. Cohen, S. Litsyn and H. Mattson extend the results of a previous paper to the nonlinear case: they investigate perfect weighted coverings (PWC) with diameter 1, finding a partial characterization, and complete an investigation on linear PWC with distance 1 and diameter 2.

G. Zémor introduces an extremal problem which can be thought of as the search for the parameters of binary linear codes of minimal distance 3 and the largest possible covering radius. A construction is proposed, lower and upper bounds are obtained that meet for certain values of the parameters.

G. Katsman presents new upper and lower bounds for the covering radius of codes which are dual to the product of parity-check codes.

Constructions

V. Pless describes greedy codes, a class of linear codes which contains lexicodes. She gives a recursive algorithm for constructing a parity-check matrix for specific greedy codes, the dimension of which can therefore be derived for some values of n and d.

I. Dumer constructs linear codes over $GF(q)$ with fixed distance and small redundancy r. In particular, he gets codes with length q^m and $r \leq 2(m+1) + \lceil m/3 \rceil$.

D. Gardy and P. Solé use saddle point techniques to derive the asymptotic volume of Lee spheres for small and large alphabets. They also obtain an asymptotic relation between covering radius and dual distance for binary codes in Hamming space.

S. Kovalev constructs convolutional codes with rate $1/k$ and constraint length $3k$ over an alphabet of size 2^k. These codes have minimum number of low weight code words.

B. Arazi considers the problem of position recovery of a specified bit in a periodical binary sequence. He introduces a new class of sequences with linear recovery complexity and compares it to de Bruijn and maximal sequences.

V. Blinovsky uses methods from random coding to get existence bounds on codes correcting errors and defects. Their construction complexity is exponentially smaller than previously obtained.

Decoding

B. Kudryashov analyses the performances of block codes obtained from convolutional codes, in terms of minimum distance, error probability and asymptotical complexity of decoding.

J. Snyders describes a partial ordering of error patterns enabling him to perform maximum likelihood soft decoding by storing only minimal elements.

E. Gabidulin gives a new construction of maximal rank distance codes and proposes a new fast matrix decoding algorithm which generalizes Peterson's algorithm for BCH codes.

A. Ashikhmin and S. Litsyn offer a linear algorithm for determining the index of the maximum spectrum element in a Fourier transform, useful for bounded Euclidean distance decoding.

V. Balakirsky derives a coding theorem and its converse for a binary-input memoryless channel with arbitrary additive decoding decision function.

I. Bocharova proposes a new approach to choosing codes and user energy for the two-user multiple-access channel.

We would like to thank the referees: G. Battail, G. Cohen, S. Litsyn, A. Lobstein, M. Perret, P. Solé, J.-P. Tillich, G. Zémor. We also thank Nathalie Le Ruyet for help in editing.

November 1991

Gérard Cohen
Simon Litsyn
Antoine Lobstein
Gilles Zémor

Contents

ON THE CORRELATION OF SEQUENCES

A. Tietäväinen

Department of Mathematics
University of Turku
SF-20500 Turku, Finland

Abstract

In code division multiple access applications it is necessary to find sequences with small maximum nontrivial correlation $C_{\max}$. In this paper we show that the moduli of certain character sums can be calculated in a very easy way by using a modification of a method of Sidelnikov. These results yield a very simple way to calculate $C_{\max}$ for nonbinary sequences found by Kumar and Moreno and for those found by Liu and Komo.

1 Introduction

Consider a family $S = \{\underline{s}_i = (s_i(t))_{t=0}^{L-1} : 1 \leq i \leq M\}$ of M sequences, each of period L, over the field F_p. Let ω be a complex primitive p^{th} root of unity. The correlation function C_{ij} between the ith and the jth sequence is defined by

$$C_{ij}(\tau) = \sum_{t=0}^{L-1} \omega^{\underline{s}_i(t \oplus \tau) - \underline{s}_j(t)}, \quad 0 \leq \tau \leq L-1$$

where $\oplus$ means addition modulo L. In code division multiple access applications the designer tries to minimize the number

$$C_{\max} = \max\{|C_{ij}(\tau)| : 1 \leq i,j \leq M\,,\ 0 \leq \tau \leq L-1;\ i \neq j \text{ if } \tau = 0\}.$$

The results of Welch [7] and Sidelnikov [4] give lower bounds for $C_{\max}$ (cf. also [6] and [1]). The Sidelnikov bound implies (see [2])

$$C_{\max}^2 > \begin{cases} L - \frac{L}{M} & \text{if } p = 2 \\ L - \frac{1}{2} - \frac{L}{M} & \text{if } p > 2 \end{cases} \tag{1}$$

and

$$C_{\max}^2 > \begin{cases} 3L - 2 - \frac{L^2}{M} & \text{if } p = 2 \\ 2L - 1 - \frac{L^2}{M} & \text{if } p > 2 . \end{cases} \tag{2}$$

Let $M \sim L^u$ when $L \to \infty$. If $u = \frac{1}{2}$, the inequality (1) implies

$$C_{\max}^2 \gtrsim L \text{ for all } p .$$

On the other hand, if $u = 1$, the inequality (2) shows that

$$C_{\max}^2 \gtrsim \begin{cases} 2L & \text{if } p = 2 \quad (3) \\ L & \text{if } p > 2 \quad (4) \end{cases}$$

The inequalities (3) and (4) seem to suggest that the nonbinary alphabets might yield better sequences than the binary ones do.

Kumar and Moreno [2] proved that this suggestion is correct by constructing a new set of nonbinary sequences which shows that the bound (4) is tight. They also determined the precise distribution of correlation values. In this paper we show that if we are not interested in the distribution of correlation values but only in $C_{\max}$, we have a very simple and elementary way to calculate it. This alternative approach, which might be suitable for some lecture courses, also gives an explanation of the better performance of nonbinary alphabets in case $u = 1$. Finally, we use the same approach in order to calculate the number $C_{\max}$ for small sets of Kasami sequences and see that we get the same result for both binary and nonbinary alphabets.

It is worth mentioning that already in [5] Solé constructed a family of quadriphase sequences for which the inequality (4) is tight.

2 Sequences and character sums

Consider the finite field F_q where $q = p^n$ and p is a prime. Let ω be a complex primitive p^{th} root of unity. Denote the trace mapping from F_q onto F_p by Tr. Then the function e defined by

$$\forall \beta \in F_q : e(\beta) = \omega^{Tr(\beta)}$$

is a nontrivial additive character of F_q. Thus

$$\sum_{x \in F_q} e(xy) = \begin{cases} q & \text{if } y = 0 \\ 0 & \text{otherwise.} \end{cases} \tag{5}$$

Assume further that $1 \le m \le n$ and define for all elements β and γ of F_q the character sum $\sigma(\beta,\gamma)$ by the equation

$$\sigma(\beta,\gamma) = \sum_{x \in F_q} e(\beta x^{p^m+1} + \gamma x).$$

Let $F = \{\beta_1, \beta_2, ..., \beta_M\}$ be a subset and α a primitive element of F_q. Put $S = \{\underline{s}_i : 1 \le i \le M\}$ where $\underline{s}_i = (s_i(t))_{t=0}^{q-2}$ is defined by

$$s_i(t) = Tr(\beta_i \alpha^{(p^m+1)t} + \alpha^t),\ 0 \le t \le q-2.$$

Then (see[2, Equations (13)–(15)])

$$C_{ij}(\tau) = \sigma(\beta_i \alpha^{(p^m+1)\tau} - \beta_j\,,\, \alpha^\tau - 1) - 1. \tag{6}$$

Thus the problem of finding $C_{\max}$ is now in a purely character sum theoretical form.

3 Kumar–Moreno sequences

Assume that $q = p^n$, $F = F_q$ and $1 \le m \le n$. If we denote $\gcd(m,n)$ by r and assume that $\frac{n}{r}$ is odd and $p > 2$, then the set S defined in Section 2 is the new family found by Kumar and Moreno [2]. By (6), the result

$$C_{\max} \le \sqrt{q} + 1$$

follows from the following theorem which we shall prove in an easy way.

Theorem 1. Let $p > 2$. Assume that $1 \le m \le n$, denote $\gcd(m,n)$ by r and assume that $e := \frac{n}{r}$ is odd. If $(\beta,\gamma) \neq (0,0)$ then

$$|\sigma(\beta,\gamma)| \le \sqrt{q}.$$

Proof. If $\beta = 0$ then $\gamma \neq 0$ and clearly $\sigma(\beta,\gamma) = 0$. Thus assume that $\beta \neq 0$. When we put $y = x + z$ and observe that

$$(x+z)^{p^m+1} = x^{p^m+1} + x^{p^m}z + xz^{p^m} + z^{p^m+1},$$

we obtain

$$\begin{aligned}|\sigma(\beta,\gamma)|^2 &= \overline{\sigma(\beta,\gamma)}\cdot\sigma(\beta,\gamma)\\ &= \sum_{x\in F_q} e(-\beta x^{p^m+1}-\gamma x)\sum_{y\in F_q} e(\beta y^{p^m+1}+\gamma y)\\ &= \sum_{x\in F_q}\sum_{z\in F_q} e(\beta x^{p^m}z+\beta x z^{p^m}+\beta z^{p^m+1}+\gamma z).\end{aligned}$$

Put $x^{p^m} = u$. Since

$$e(\beta x z^{p^m}) = e(\beta^{p^m} x^{p^m} z^{p^{2m}}),$$

the above equation implies

$$|\sigma(\beta,\gamma)|^2 = \sum_{z\in F_q} e(\beta z^{p^m+1}+\gamma z)\sum_{u\in F_q} e(u\beta z(\beta^{p^m-1} z^{p^{2m}-1}+1)).$$

By (5), the inner sum is nonzero (and then equal to q) if and only if $z = 0$ or

$$\beta^{p^m-1} z^{p^{2m}-1} = -1. \tag{7}$$

The logarithm of the left hand side of (7) is divisible by $p^r - 1$. On the other hand, the logarithm of the right hand side equals

$$\frac{p^n-1}{2} = \frac{(p^r-1)(1+p^r+p^{2r}+\ldots+p^{(e-1)r})}{2}$$

and therefore it is not divisible by p^r-1. Hence Equation (7) is not solvable in F_q and thus $|\sigma(\beta,\gamma)|^2 = q$ □

Remark. If $p = 2$, the logarithm of the right hand side of (7) is zero and therefore we can find β's such that Equation (7) is solvable in F_q. Thus $|\sigma(\beta,\gamma)|^2$ may be and in fact often is greater than q. This difference between binary and nonbinary character sums gives the advantage the Kumar–Moreno sequences have compared with the corresponding binary sequences.

4 Kasami sequences

Assume that $q = p^n$, $n = 2m$, and ε is an element of F_q such that $\varepsilon^{p^m}+\varepsilon \neq 0$. Let $F = \{\beta_i = \varepsilon\delta_i : 1 \leq i \leq p^m\}$ where δ_i takes on each value of F_{p^m} when i runs from 1 to p^m. Now the sets S defined in Section 2 are small sets of Kasami sequences (for the nonbinary case, see [3]). Using Equation (6) and the definition of $C_{\max}$ we infer that

$$C_{\max} \leq \max\{|\sigma(\beta,\gamma)| : \beta = \varepsilon\delta\,,\ \varepsilon^{p^m}+\varepsilon \neq 0\,,\ \delta \in F_{p^m}\,,\ (\delta,\gamma)\neq(0,0)\} + 1\,.$$

Consequently, the result

$$C_{\max} \leq \sqrt{q}+1$$

can be proved by proving the following theorem.

Theorem 2. If $n = 2m$, $\beta = \varepsilon\delta$, $\varepsilon^{p^m}+\varepsilon \neq 0$, $\delta \in F_{p^m}$ and $(\delta,\gamma) \neq (0,0)$ then

$$|\sigma(\beta,\gamma)| \leq \sqrt{q}\,.$$

Proof. If $\delta = 0$ then $\gamma \neq 0$ and clearly $\sigma(0,\gamma) = 0$. Thus assume $\delta \neq 0$. When we put $y = x + z$ and observe that

$$e(\beta x^{p^m} z) = e(\beta^{p^m} x z^{p^m}),$$

we obtain

$$\begin{aligned}|\sigma(\beta,\gamma)|^2 &= \sum_{x\in F_q}\sum_{z\in F_q} e(\beta x^{p^m}z+\beta x z^{p^m}+\beta z^{p^m+1}+\gamma z)\\ &= \sum_{z\in F_q} e(\beta z^{p^m+1}+\gamma z)\sum_{x\in F_q} e(x(\varepsilon^{p^m}+\varepsilon)\delta z^{p^m}).\end{aligned}$$

Since $\delta \neq 0$ and $\varepsilon^{p^m}+\varepsilon \neq 0$, the inner sum is nonzero (and then equal to q) if and only if $z = 0$. Thus $|\sigma(\beta,\gamma)|^2 = q$ □

Hence in this case we have the same result for both binary and nonbinary alphabets.

References

[1] P. Vijay Kumar and Chao-Ming Liu: On lower bounds to the maximum correlation of complex roots-of-unity sequences. — IEEE Transactions on Information Theory 36 (1990), 633–640.

[2] P. Vijay Kumar and Oscar Moreno: Polyphase sequences with periodic correlation properties better than binary sequences. — IEEE Transactions on Information Theory (submitted).

[3] Shyh-Chang Liu and John J. Komo: Nonbinary Kasami sequences over GF(p). — IEEE Transactions on Information Theory (submitted).

[4] V.M. Sidelnikov: On mutual correlation of sequences. — Problemy Kibernetiki 24 (1971), 15–42 (in Russian).

[5] P. Solé: A quaternary cyclic code and a family of quadriphase sequences with low correlation properties. — Coding Theory and Applications (Eds. G. Cohen and J. Wolfmann), Lecture Notes in Computer Science 388, Springer-Verlag (1989), 193–201.

[6] Hannu Tarnanen: On character sums and codes. — Discrete Mathematics 57 (1985), 285–295.

[7] L.R. Welch: Lower bounds on the maximum correlation of signals. — IEEE Transactions on Information Theory 20 (1974), 397–399.

ON GAUSSIAN SUMS FOR FINITE FIELDS AND ELLIPTIC CURVES

Shparlinski I.E.

In this paper we obtain a new bound for Gaussian sums

$$G_n(a,\ p) = \sum_{x\in\mathbb{F}_p} \exp(2\pi i a x^n/p), \quad (a,\ p) = 1,$$

in prime finite fields $\mathbb{F}_p$ of p elements that improves the well-known classical bound $|G_n(a,\ p)| \leq np^{1/2}$ (see [18], Ch.5) and is non-trivial for

$$2^{-12/7}p^{4/7} > n > 2^{12/5}p^{2/5}.$$

Then, using the method of the papers [24], [25], we prove a nontrivial upper bound for Gaussian sums in finite fields $\mathbb{F}_{2^m}$ of 2^m elements.

Finally, we obtain upper bounds for an analogy of Gaussian sums for points of elliptic curves $\mathbb{E}_p$ over $\mathbb{F}_p$ and give their applications to the distribution of primitive points of elliptic curves with a cyclic group of points (i.e. their generating points). Such points were considered in for the first time [17]. Interest for these points has been growing for the last few years, because of their applications to cryptography (see [7], [8], [10], [13], [14]).

Let us define

$$G_n(p) = \max_{a\in\mathbb{F}_p^*} |G_n(a,\ p)|.$$

Theorem 1. $G_n(p) \leq 2n^{7/12}p^{2/3}$.

Proof. It is enough to prove the bound for $n \mid (p - 1)$ because of

$$G_n(a,\ p) = G_{(n,p-1)}(a,\ p)$$

and $n \geq (p - 1)^{1/4} + 1$ (for smaller values of n the $G_n(p) \leq np^{1/2}$ is better).

After some evaluation, it can be verified, that

$$\sum_{a\in\mathbb{F}_p} |G_n(a, p)|^4 = p \sum_{a\in\mathbb{F}_p} N^2_{n,p}(a).$$

where $N_{n,p}(a)$ is the number of rational points on the Fermat curves $x^n + y^n = a$, $x, y \in \mathbb{F}_p$

It follows from [6] that for $a \neq 0$ and $n > (p - 1)^{1/4} + 1$ the bound

$$N_{n,p}(a) \leq 4n^{4/3}(p - 1)^{2/3}$$

holds. Using this bound, the trivial bound $N_{n,p}(0) \leq np$, and the equality

$$\sum_{a\in\mathbb{F}_p} N_{n,p}(a) = p^2,$$

we obtain

$$\sum_{a\in\mathbb{F}_p} |G_n(a, p)|^4 \leq 5n^{4/3}(p - 1)^{2/3}p^3.$$

Taking into account that $G_n(a, p) = G_n(ab^n, p)$, we get the bound. 4

Th.1 implies a positive answer to the question stated in [30] on a uniform bound of a constant in the bound for Gaussian sums with an arbitrary composite denominator.

Corollary 1.

$$\max_{n\geq 2} \max_{m\in\mathbb{N}} \max_{(a,m)=1} m^{-1+1/n}\left|\sum_{x=1}^{m} \exp(2\pi iax^n/m)\right| < \infty$$

Note that it would be interesting to compute the value (in the right-hand side) exactly and to determine for which n, m and a this value is attained. The method of [30] and Th.1 allow this.

Other consequences of Th.1 are the improvements of the bound of [15] of exponential sums with exponential functions and of the result of [21] on the rank of the Hasse-Witt matrix of hyperelliptic functions over finite fields.

Let g be an integer, $(g, p) = 1$, and let t_p be the exponent of g modulo p. It is known (see [15], [20]) that the following bound of the exponential sum

(1) $$\left|\sum_{z=1}^{t_p} \exp[2\pi i(ag^z/p + bz/t_p)]\right| < p^{1/2},$$

holds, where a and b are integers, $(a, p) = 1$.

Let $\mathbf{ind}z$ be the discrete logarithm modulo p of z with respect to some fixed primitive root ϑ.

Setting $n = (p - 1)/t_p$, $\chi(x) = \exp[2\pi i b \mathbf{ind}x/(p - 1)]$, we obtain from Th.1

(2) $$\left|\sum_{z=1}^{t_p} \exp[2\pi i(ag^z/p + bz/t_p)]\right| < 3t_p^{5/12}p^{1/4}.$$

This bound improves (1) and the trivial bound for t_p in the interval

$$3^{-12/5}p^{3/5} > t_p > 3^{12/7}p^{3/7}.$$

Note that for any fixed $g > 1$ and all prime $p \leqslant x$, $p \nmid g$, except possibly $o(\pi(x))$ primes, we have $t_p > p^{1/2-\varepsilon}$. In fact, if $Q(x)$ is the set of these exclusive primes then any $p \in Q(x)$ divides the product

$$R(x) = \prod_{t < x^{1/2-\varepsilon}} (g^t - 1).$$

Then, $|Q(x)| \leqslant \nu(R(x))$, where $\nu(k)$ is the number of prime divisors of natural k. Using the following well known estimate

$$\nu(k) = O(\log k/\log\log k)$$

and the bound $\log R(x) = O(x^{1-2\varepsilon})$. Hence $|Q(x)| = o(\pi(x))$.

Therefore, the bound (2) for any fixed $g > 1$ and almost all $p \in \mathbb{P}$ gives

$$\left|\sum_{z=1}^{t_p} \exp[2\pi i(ag^z/p + bz/t_p)]\right| \leqslant t_p^{11/12+\varepsilon}.$$

In [21] the following interesting problem on congruences was considered. Let $p \in \mathbb{P}$ and $T(p)$ be the largest value of all such natural t for which there exists g, (g, p) = 1, with the exponent t modulo p and the property that all residues (mod p) of g^x, $x = 1, 2, \dots, t$, belong to the interval $[1, (p-1)/2]$. In that paper it was proved that $T(p) = O(p^{1/2}\log p)$. We stated the following improvement based on the bound (2) (see [28]).

Corollary 2. $T(p) = O(p^{3/7})$.

Other consequences of Th.1 are given in [28].

For $z \in \mathbb{F}_{2^m}$ define by $\mathrm{Sp}(z)$ its trace in $\mathbb{F}_2$, and let

$$\psi(z) = (-1)^{\mathrm{Sp}(z)}.$$

Let us consider the sums

$$S_{m,n}(a) = \sum_{z\in\mathbb{F}_{2^m}} \psi(az^n), \quad a \in \mathbb{F}_{2^m},$$

and set

$$S_{m,n} = \max_{a\in\mathbb{F}^*_{2^m}} |S_{m,n}(a)|.$$

Of course, we have $S_{m,n} \leq n2^{m/2}$ (see [18], Ch.5). Below we state a bound that is non-trivial for $n > 2^{m/2}$.

Let $\vartheta(\alpha)$ be the root of the equation

$$\vartheta\log\vartheta + (1-\vartheta)\log(1-\vartheta) = (\alpha - 1)\log 2, \quad 0 < \vartheta < 1/2,$$

(it is easy to see that this equation has a unique root for $0 < \alpha < 1$).

Set $\gamma(\alpha) = 1 - \vartheta(\alpha)$.

Theorem 2. For any fixed α, $0 < \alpha < 1$, $n < 2^{\alpha m}$ **with**

$$n(2^k - 1) \not\equiv 0 \pmod{2^m - 1}, \quad k = 0, 1, \dots, m-1,$$

the bound $S_{m,n} < \gamma(\alpha)2^m + o(2^m)$, $m \to \infty$, **holds.**

Proof. We can suppose that $n \mid 2^m - 1$. Let g be a primitive root of $\mathbb{F}_{2^m}$. Then for $a \in \mathbb{F}^*_{2^m}$ we have

$$S_{m,n}(a) = 1 + n \sum_{x=1}^{\tau} (-1)^{u(x)}$$

where $u(x) = \mathrm{Sp}(ag^{nx})$, $\tau = (2^m - 1)/n$. Therefore

$$S_{m,n}(a) = 1 + n(\tau - 2W),$$

where W is the number of solutions of the equation $u(x) = 1$, $1 \leqslant x \leqslant \tau$.

Note that $u(x)$ is a linear recurring sequence of order m and with minimal period $\tau \geqslant 2^{(1-\alpha)m}$ (see [18], Th.8.24).

Applying Th.2 of [24] we get $W > \vartheta(\alpha)\tau + o(\tau)$. Analogously it can be proved that $W < (1 - \vartheta(\alpha))\tau + o(\tau)$ and we have the result. 4

Denote by $G(m,n)$ the smallest r such that the equation

$$x_1^n + \ldots + x_r^n = c, \quad x_1, \ldots, x_r \in \mathbb{F}_{2^m},$$

is solvable for any $c \in \mathbb{F}_{2^m}$.

Corollary. For any fixed α, $0 < \alpha < 1$, $n < 2^{\alpha m}$ **with**

$$n(2^k - 1) \not\equiv 0 \pmod{2^m - 1}, \quad k = 0, 1, \ldots, m - 1,$$

the bound $G(m,n) = O(\log n)$ **holds.**

Now we are going to consider elliptic curves (see [12], [16], [29], [31]).

These curves appear in many different areas of mathematics and computer science such as primality testing and integer factoring algorithms, coding theory, cryptography , etc (see [1], [2], [4], [5], [10], [11], [13], [14], [19], [22], [27] and references there).

All applications are based on the fact that an elliptic curve $\mathbb{E}$ over any field can be considered as an Abelian group under an appropriate composition rule and with some "point at infinity" I as the unit (we write the group law multiplicatively, the sign "+" we will keep for the usual vector addition).

It is known that in any finite field $\mathbb{F}_q$ of characteristic $p > 3$ (and more generally, in an arbitrary field $\mathbb{F}$ of characteristic $p \neq 2$,

3) any elliptic curve has an affine model given by the Weierstrass equation

(3) $$y^2 = x^3 + Ax + B, \quad A,\ B \in \mathbb{F}_q,$$

with the discriminant

$$\Delta = 4A^3 + 27B^2 \neq 0.$$

It is not difficult to point out an elliptic curve with a cyclic group of points. For example, if $p \equiv 1 \pmod 4$ then the curve given by the Weierstrass equation $y^2 = x^3 + x$, has a cyclic group of points over $\mathbb{F}_p$ (see Th.6 of [11]).

In [9] the distribution of $p \in \mathbb{P}$, for which the reduction (mod p) of a given elliptic curve $\mathbb{E}$ over $\mathbb{Q}$ has a cyclic group of points, was studied. Under some simple assumptions about $\mathbb{E}$, for the number $T_{\mathbb{E}}(x)$ of such primes $p < x$, the lower bound $T_{\mathbb{E}}(x) > cx/\log^2 x$ was established, where $c > 0$ is some constant. The reduction (mod p) of free subgroups of $\mathbb{E}$ was treated also (under the Extended Riemann Hypothesis).

For a given elliptic curve over $\mathbb{F}_p$ with a cyclic group of points we establish below some results on the distribution of primitive points that allow to find such a point in time $O(p^{2/3+\varepsilon})$ (by the "brute-force" searching over all $x = 1, \ldots, h$ for certain $h = O(p^{2/3+\varepsilon})$).

For $(a,b) \in \mathbb{F}_p^2 \setminus \{(0,0)\}$ let us define the additive character $\psi(P) = \exp[2\pi i(ax + by)/p]$ on points $P = (x,y) \in \mathbb{F}_p^2$ and let us consider sums, analogous to Gaussian sums, for the affine elliptic curve $\mathbb{E}_p$ given by the Weierstrass equation (3):

$$\sigma_m(\psi) = \sum_{P \in \mathbb{E}_p} \psi(P^m)$$

(for the point at infinity we set $\psi(I) = 1$). Using the bound of exponential sums along curves (see [3]) and the explicit formulas for the coordinates of P^m with respect to the coordinates of a point $P \in \mathbb{E}$ (see [12], [16], [29], [31]), for nontrivial character ψ we easily obtain (see [26])

(4) $$\sigma_m(\psi) = O(m^2 p^{1/2}).$$

Moreover, for the sum over points of any subgroup $\mathbb{H} \subseteq \mathbb{E}_p$,

$$S_{\mathbb{H}}(\psi) = \sum_{P\in\mathbb{H}} \psi(P),$$

we have

$$|\mathbb{H}|\,|S_{\mathbb{H}}(\psi)|^2 = \sum_{Q\in\mathbb{H}} |\sum_{P\in\mathbb{H}} \psi(QP)|^2 \leq \sum_{Q\in\mathbb{E}_p} |\sum_{P\in\mathbb{H}} \psi(QP)|^2$$

$$= \sum_{P_1\in\mathbb{H}} \sum_{P_2\in\mathbb{H}} \sum_{Q\in\mathbb{E}_p} \psi(QP_1)\overline{\psi(QP_2)} = |\mathbb{H}| \sum_{P\in\mathbb{H}} \sum_{Q\in\mathbb{E}_p} \psi(QP)\overline{\psi(Q)}.$$

For each $P \in \mathbb{H}$, $P \neq I$, and for $Q = (x, y) \in \mathbb{E}_p$, $Q \neq I$, $Q \neq P^{-1}$ we have

$$\psi(QP)\overline{\psi(Q)} = \mathbf{exp}(2\pi i R_p(x, y)/p),$$

where $R_p(x, y)$ is a rational function of degree $O(1)$ with coefficients depending on P. Writing this function in an explicit form, it is not difficult to prove that $R_p(x, y)$ is a constant along $\mathbb{E}_p$ for at most $O(1)$ values of P. For these P we can apply the bound of [3] to the sum over Q. In the opposite case we bound this sum as $O(p)$.

Therefore

$$|S_{\mathbb{H}}(\psi)|^2 \leq O(|\mathbb{H}|p^{1/2} + p).$$

Since for $|S_{\mathbb{H}}(\psi)| \leq |\mathbb{H}|$ and this bound is trivial, we can suppose that $|\mathbb{H}| \geq p^{1/2}$, then

$$S_{\mathbb{H}}(\psi) = O\left(|\mathbb{H}|^{1/2}p^{1/4}\right). \tag{5}$$

If $\mathbb{E}_p$ has a cyclic group of points then sums $\sigma_m(\psi)$ and $S_{\mathbb{H}}(\psi)$ can be reduced to each other as follows

$$\sigma_m(\psi) = (m, N)S_{\mathbb{H}_m}(\psi),$$

where $\mathbb{H}_m = \{P^m \mid P \in \mathbb{E}_p\}$. Since $|\mathbb{H}_m| = N/(m, N)$ and $N = O(p)$, we obtain:

Theorem 3. If $\mathbb{E}_p$ is an elliptic curve with a cyclic group of points then for a nontrivial character ψ

$$\sigma_m(\psi) = O(\min(m^2 p^{1/2},\ m^{1/2} p^{3/4})).$$

Th.3 implies that

$$\sigma_m^3(\psi) = O\big(m^2 p^{1/2} (m^{1/2} p^{3/4})^2\big) = O(m^3 p^2),$$

$$\sigma_m(\psi) = O(mp^{2/3}).$$

Using this bound and applying the same arguments as in Th.6 of [26] (where the weaker bound $\sigma_m(\psi) = O(mp^{3/4})$ was used) one can obtain the asymptotical formula

$$G(\mathfrak{B}) = |\mathfrak{B}|\varphi(N)/p^2 + O(p^{2/3+\varepsilon}),$$

where $G(\mathfrak{B})$ is the number of primitive points of a given elliptic curve with a cyclic group of points which is contained in the box

$$\mathfrak{B} = [0,\ h-1] \times [0,\ k-1],$$

and N is the number of all $\mathbb{F}_p$-rational points over $\mathbb{E}_p$. This formula improves the analogous result of [26] where only the bound (4) was used.

It follows from the last asymptotical formula and well-known results on solving quadratic congruences (see [1]), that a primitive point of a given elliptic curve with a cyclic group of points can be found using $O(p^{2/3+\varepsilon})$ arithmetical operations in $\mathbb{F}_p$. Powering this point we obtain all points of $\mathbb{E}_p$ using $O(p)$ arithmetical operations in $\mathbb{F}_p$.

These considerations imply

Theorem 4. All $\mathbb{F}_p$-rational points of an elliptic curve $\mathbb{E}_p$ given by the Weierstrass equation (3) and with a cyclic group of points can be found using $O(p)$ arithmetical operations in $\mathbb{F}_p$.

It is clear, that this bound cannot be improved because $N \sim p$. The direct search of all points of $\mathbb{E}_p$ uses $p\log^c p$, $c > 1$, arithmetical operations in $\mathbb{F}_p$. It is related to the complexity of solving square congruences (see [22], [23]).

The list of all $\mathbb{F}_p$-rational points of an elliptic curve $\mathbb{E}_p$ is required when constructing algebraic geometric codes over these curves (see [2], [4], [5], [11]).

We present a direct consequence of the bound (5) for the reduction (**mod** p) of a free subgroup of a given elliptic curve $\mathbb{E}$ over $\mathbb{Q}$.

Let $\mathbb{E}$ be an elliptic curve over $\mathbb{Q}$ with the rank $r_{\mathbb{E}}$. Let us suppose also that we have an independent set of $r \leq r_{\mathbb{E}}$ rational points of $\mathbb{E}$. Let Γ be generated by these points and Γ_p be the reduction of Γ modulo a prime p.

Theorem 5. If $r \geq 3$ then for all except possibly $o(\pi(x))$ prime numbers $p \leq x$, we have the bound

$$S_{\Gamma_p}(\psi) = O\left(|\Gamma_p| p^{-(r-2)/4(r+2)+\varepsilon}\right).$$

Proof. It was proved in [9] (see also [19]) that the number of primes p for which $|\Gamma_p| < y$ is $O(y^{1+2/r})$. Then for all, except possibly $o(\pi(x))$, prime numbers $p \leq x$ the bound $|\Gamma_p| \geq p^{r/(r+2)-\varepsilon}$ holds. For these primes we have from (5)

$$S_{\Gamma_p}(\psi)/|\Gamma_p| = O\left(|\Gamma_p|^{-1/2} p^{1/4}\right) = O\left(p^{-(r-2)/4(r+2)+\varepsilon}\right),$$

the theorem is proved. 4

The following estimate can be derived quite analogously to (5).

$$\sum_{P \in \mathbb{E}_p} \chi(P)\psi(P) = O(p^{3/4})$$

where χ is a nontrivial character of the multiplicative group of points over $\mathbb{E}_p$.

REFERENCES

1. BACH E. Number-theoretic algorithms. Annual Review of Comp. Sci., 1990, v.4, p.119-172.
2. BARG F.V., KATSMAN G.S., TSFASMAN M.A. Algebraic-geometric codes over curves of small genus. Problemy Peredachi Inform., 1987, v.23, no.1, p.42-46 (in Russian).
3. BOMBIERI E. On exponential sums in finite fields. Amer. J. Math., 1966, v.88, p.71-105.
4. DRIENCOURT Y. Some properties of elliptic codes over a field of characteristic 2. Lecture Notes in Comp. Sci., 1985, v.229.
5. DRIENCOURT Y., MICHON J.F. Elliptic codes over fields of characteristic 2. J. Pure and Appl. Algebra, 1987, v.45, p.15-39.
6. GARCIA A., VOLOCH J.F. Fermat curves over finite fields. J. Number Theory. 1988, v.30, p.345-356.
7. GUPTA R. Division fields of $Y^2 = X^3 - aX$. J. Number Theory, 1990, v.34, no.3, p.335-345.
8. GUPTA R., MURTY M.R.P. Primitive points on elliptic curves. Compos. Math., 1986, v.58, no.1, p.13-44.
9. GUPTA R., MURTY M.R.P. Cyclicity and generation of points mod p on elliptic curves. Invent. Math., 1990, v.111, no.1, p.225-235.
10. KALINSKI B.S. A pseudo-random bit generator based on elliptic logarithm. Lect. Notes in Comp. Sci., 1987, v.263, p.84-103.
11. KATSMAN G.L., TSFASMAN M.A. Spectrums of algebraic-geometric codes. Problemy Peredachi Inform. 1987, v.23, no.4, p.19-34 (in Russian).
12. KOBLITZ N. Introduction to elliptic curves and modular forms. Springer-Verlag, 1984.
13. KOBLITZ N. Elliptic curve cryptosystem. Math. Comp., 1987, v.48, no.177, p.203-209.
14. KOBLITZ N. A course in number theory and cryptography. Springer-Verlag, 1987.
15. KOROBOV N.M. On the distribution of digits in periodic fractions. Matem. Sbornik, 1972, v.89, no.4, p.654-670 (in Russian).
16. LANG S. Elliptic curves: Diophantine analysis. Springer-Verlag, 1978.
17. LANG S., TROTTER H. Primitive points on elliptic curves. Bull. Amer. Math. Soc., 1977, v.83, p.289-292.
18. LIDL R., NIEDERREITER H. Finite fields. Addison-Wesley, 1983.
19. MIYAMOTO I., MURTY M.R.P. Elliptic pseudoprimes. Math. Comp., 1989, v.53, no.187, p.415-430.
20. NIEDERREITER H. Quasy-Monte Carlo methods and pseudo-random numbers. Bull. Amer. Math. Soc., 1978, v.84, p.957-1041.
21. NIEDERREITER H. On a problem of Kodama concerning the Hasse-Witt matrix and distribution of residues. Proc. Japan Acad., Ser.A, 1987, v.63, no.9, p.367-369.
22. SCHOOF R.J. Elliptic curves over finite fields and the computation of square roots mod p. Math. of Comp., 1985, v.44, no.170, p.483-494.
23. SHANKS D. Five number theoretic algorithms. Proc. 2 Manitoba Conf. on Numerical Math., Univ. Manitoba, 1972, p.51-70.
24. SHPARLINSKI I.E. On some properties of linear cyclic codes. Problemy Peredachi Inform., 1983, v.19, no.3, p.106-110 (in Russian).

25. SHPARLINSKI I.E. On weigth enumerators of some codes. Problemy Peredachi Inform., 1986, v.22, no.2, p.43-48 (in Russian).
26. SHPARLINSKI I.E. On primitive elements in finite fields and on elliptic curves. Matem. Sbornik, 1990, v.181, no.9, p.1196-1206 (in Russian).
27. SHPARLINSKI I.E. On some problems of theory of finite fields. Uspechi Matem. Nauk, 1991, v.46, no.1, p. 165-200 (in Russian)
28. SHPARLINSKI I.E. On bounds of Gaussian sums. Matem. Zametki, 1991, (to appear).
29. SILVERMAN J.H. The arithmetic of elliptic curves. Springer-Verlag, 1984.
30. STECHKIN S.B. A bound of sums of Gauss. Matem. Zametki, 1975, v.17, no.4. p.579-588 (in Russian).
31. TATE J. The arithmetic of elliptic curves. Invent. Math., 1974, v.23, p.179-206.

EXPONENTIAL SUMS AND CONSTRAINED ERROR-CORRECTING CODES

ALEXANDER BARG

Institute for Problems
of Information Transmission
Ermolovoy 19, Moscow GSP-4
101447 U.S.S.R.
E-mail: aBarg@ippi.msk.su

Abstract. We present a number of new families of k-ary dc-constrained error-correcting codes with distance $d \gtrsim (k-1)n/k - \alpha_1(n)\sqrt{n}$ and running digital sum $\cong \alpha_2(n)\sqrt{n}$, where α_1 and α_2 are slowly growing functions in the code length n. We show also that constructed codes are comma-free and detect synchronization errors even at high rate of additive errors. To prove these properties of constructed codes, we apply some well-known inequalities for incomplete sums of characters of polynomials.

1. INTRODUCTION

Throughout the paper $A[l, n, M, d]$ denotes a l-ary code of length n, with M words and minimum Hamming distance d. For p an odd prime let $q = p^s$ and denote by $\mathfrak{P} \subset \{f(x) \in \mathbb{F}_q[x] | \ 1 \leqslant \deg f \leqslant r < q-1\}$ a certain subset of the set of all monic polynomials of degree $\leqslant r$. Next, let $\chi : \mathbb{F}_q^* \to \mathbb{C}$ be a multiplicative character of order $k|(q-1)$ of the field $\mathbb{F}_q = (u_0, u_1, \ldots, u_{q-1})$. In this paper, we study codes defined as follows:

$$A = \{\mathbf{a}_f, f \in \mathfrak{P}\}, \tag{1}$$

$$\mathbf{a}_f = (\chi(f(u_0)), \chi(f(u_1)), \ldots, \chi(f(u_{q-1}))) \tag{2}$$

and related codes. After substituting a certain root of unity for $\chi(0)$ in (2), we obtain a k-ary code with the parameters $(k, n = q, |\mathfrak{P}|, d)$, where $d = \min_{\substack{\mathbf{a},\mathbf{b} \in A \\ \mathbf{a} \neq \mathbf{b}}} \operatorname{dist}(\mathbf{a}, \mathbf{b})$ and the Hamming distance between two vectors can be calculated as

$$\operatorname{dist}(\mathbf{a}, \mathbf{b}) = n - \frac{1}{k} \sum_{i=0}^{n-1} \sum_{j=1}^{k} (a_i \bar{b}_i)^j,$$

where the bar denotes complex conjugation. Generally, the distance of the code A may appear to be quite small (or even zero) if the set $\mathfrak{P}$ was chosen unsuccessfully. For a vector $\mathbf{a} = (a_0, a_1, \ldots, a_{n-1})$ define its running digital sum (RDS) at a moment $v, 0 \leqslant v \leqslant n-1$, as

$$s_v(\mathbf{a}) = |\sum_{j=0}^{v} a_j|$$

and the maximum RDS as $S(\mathbf{a}) = \max_{0 \leqslant v \leqslant n-1} s_v(\mathbf{a})$. For every $\mathbf{a}$ let $s_{-1}(\mathbf{a}) \triangleq 0$. The maximum RDS of a code A is, by definition,

$$S(A) = \max_{\mathbf{a} \in A} S(\mathbf{a}). \tag{3}$$

A code A is called *balanced* if for every $\mathbf{a} \in A$ we have $s_{n-1}(\mathbf{a}) = 0$, and *dc-constrained* if

$$S(A) \leqslant c(n),$$

where $c(n)$ is a slowly growing function in n.

The aim of the first part of the present paper (Section 2) is to study parameters of dc-constrained error-correcting codes defined by (1)-(2) for a specific choice of k, q, and $\mathfrak{P}$. In this part, we extend the results of [3] from prime to composite fields and from binary to arbitrary alphabet. It is evident that any code becomes balanced if one adds a tail of appropriate length and weight to each codeword. Therefore, the main restriction that forms the difference between our problem and conventional coding-theoretic problems is that on the value of RDS.

In Section 3, we consider construction of codes that maintain synchronization. The definitions that follow were introduced in [6] (see also [2], [7]). For binary vectors $\mathbf{x} = (x_0, x_1, \ldots, x_{n-1})$ and $\mathbf{y} = (y_0, y_1, \ldots, y_{n-1})$ define an n-vector $T_i(\mathbf{x}, \mathbf{y})$ as follows:

$$T_i(\mathbf{x}, \mathbf{y}) = (x_i, x_{i+1}, \ldots, x_{n-1}, y_0, y_1, \ldots, y_{i-1}), \ 1 \leqslant i \leqslant n-1.$$

For a code A define the *code separation* as

$$\rho(A) = \min_{\substack{1 \leqslant i \leqslant n-1 \\ \mathbf{a},\mathbf{b},\mathbf{c} \in A}} \operatorname{dist}(T_i(\mathbf{a}, \mathbf{b}), \mathbf{c}), \tag{4}$$

where $\mathbf{a}, \mathbf{b}$, and $\mathbf{c}$ are not necessarily different. Codes with $\rho(A) > 0$ are called *comma-free* [5].

In Section 3, we show that some error-correcting dc-constrained codes of the type (1)-(2) are comma-free and estimate their code separation.

2. List of dc-constrained codes

Here we list code families that arise from the construction (1)-(2) after a particular choice of $\mathfrak{P}, q$, and k.

2.1. *Binary codes* ($k = 2$).

For a set of binary vectors W denote by $\overline{W}$ the set $\{-\mathbf{a} | \ \mathbf{a} \in W\}$. Let $q = p^s$ be a prime or a prime power, p-odd. In binary case $\chi(c)$ is a quadratic character of $\mathbb{F}_q$ equal to 1 or to -1 depending on whether the equation $x^2 = c$ has (resp., has not) a solution in $\mathbb{F}_q$.

2.1.1. Let q be a prime of the form $4m - 1$ and choose $\mathfrak{P} = \{x + c | \ c \in \mathbb{F}_p\}$, and $\chi(0) = -1$. The code $A = \{(1|\mathbf{a}_f),\ f \in \mathfrak{P}\}$, where $\mathbf{a}_f$ is defined by (2), obviously is balanced since exactly half of nonzero elements of $\mathbb{F}_q$ are quadratic residues. The balanced code

$$B = A \bigcup \overline{A}$$

has twice as many words as A, same minimum distance and is in fact the Hadamard-Paley code (without all-one and all-minus one vectors) [8, ch. 2]. For every $\mathbf{b} \in B$ its RDS may be estimated by the Vinogradov-Polya inequality for incomplete exponential sums in $\mathbb{F}_q$ [10, p. 84]:

$$(5) \qquad |\sum_{x=1}^{b} \chi(x)| = |\sum_{x=h+1}^{h+b} \chi(x)| < \sqrt{q} \log q, \quad b < q.$$

We conclude that the balanced dc-constrained code B has the parameters

$$[2, q+1, 2q, (q+1)/2], \quad S(B) < \sqrt{q} \log q + 1.$$

2.1.2. $q = 4m + 1$ – a prime, $\mathfrak{P}$ same as above. Following [1], consider a code A (1)-(2) and set $\chi(0) = 1$. Add to A a one-bit tail to obtain the balanced code A^* with the parameters

$$[2, q+1, q-1, (q-1)/2], \quad S(A^*) < \sqrt{q} \log q + 1.$$

Next, the balanced code $B = A^* \bigcup \overline{A^*}$ has the parameters

$$[2, q+1, 2q-2, (q-5)/2], \quad S(B) < \sqrt{q} \log q + 1,$$

where the distance value comes from [1].

2.1.3. $q = p^s$, where p is an odd prime, $s \geqslant 1$, $r \lesssim \sqrt{q}/2$, and $\mathfrak{P}$ is the set of all monic non-constant square-free polynomials f (which means that the factorization of every polynomial $f \in \mathfrak{P}$ consists of different irreducible factors). Suppose that every $f \in \mathfrak{P}$ has $\deg f \leqslant r$. Order the elements of $\mathbb{F}_q$ lexicographically with respect to some basis of $\mathbb{F}_q$ over $\mathbb{F}_p$. Consider the code A (1), where for $\chi(0)$ in the lth coordinate of a codeword $\mathbf{a}, 0 \leqslant l \leqslant q-1$, we substitute -1 if $\sum_{i=0}^{l-1} a_i \geqslant 0$ and 1 if $\sum_{i=0}^{l-1} a_i < 0$. Then for every $\mathbf{a} \in A$ the value $s_{q-1}(\mathbf{a})$ is at most $(r-1)\sqrt{q}$ which is therefore the maximum length of the tail that balances A. The balanced code $B = A^* \bigcup \overline{A^*}$ has the parameters

$$\left[2, n \leqslant q + (r-1)\sqrt{q}, M = 2q^r, d \geqslant \frac{1}{2}(q - (2r-1)\sqrt{q}) - 2r\right] \tag{6}$$

and its RDS can be estimated from above by a corresponding generalization of the Vino-gradov-Polya inequality (5).

LEMMA. *Suppose $q = p^s$ is a prime power, $\chi : \mathbb{F}_q^* \to \mathbb{C}$ is a multiplicative character of order k with $\chi(0) = 0$, and $f \in \mathbb{F}_q[x]$ is a polynomial of degree r that is not equal to the kth power of another polynomial. Let $\{\alpha_1, \ldots, \alpha_s\}$ be a basis of $\mathbb{F}_q$ over $\mathbb{F}_p$. If*

$$B = \{x_1\alpha_1 + \cdots + x_s\alpha_s | \quad \forall l : \ h_l + 1 \leqslant x_l \leqslant h_l + b_l\},$$

then

$$\left|\sum_{\mathbf{x} \in B} \chi(f(\mathbf{x}))\right| < rp^{s/2}(1 + \log p)^s. \tag{7}$$

Thus we have

$$S(B) < rp^{s/2}(1 + \log p)^s. \tag{8}$$

For q a prime the codes considered here were studied in [3].

2.2. *Nonbinary codes.*

Let k divide $q-1$, where q is a prime or a prime power, $q = p^s$, p - odd.
2.2.1. Let $\mathfrak{P} = \{x + a | \ a \in \mathbb{F}_q\}$, $\chi(0) = 1$, and k even. The code A defined by (1) is a nonbinary analog of the Hadamard-Paley matrix. Since

$$\sum_{x \in \mathbb{F}_q} \chi(x + a) = 1,$$

the code becomes balanced after adding the all-minus one column. It follows from [9] that the resulting balanced code A^* has the parameters $[k, q+1, q, d]$, where

$$d = \begin{cases} \dfrac{(n-1)(k-1)}{k} & \text{if } \chi(-1) = 1; \\ \dfrac{(n-1)(k-1)}{k} + 1 & \text{if } \chi(-1) = -1. \end{cases}$$

Finally, (5) implies $S(A^*) < \sqrt{q}\log q + 1$.

2.2.2. k - even. Order elements of $\mathbb{F}_q$ as in 2.1.3. Let $\mathfrak{P}$ consist of all monic polynomials $f(x)$ of degree $1 \leqslant \deg f \leqslant r \lesssim \sqrt{q}/2$ that satisfy the two following properties:

A1. No $f \in \mathfrak{P}$ contains a factor equal to the kth degree of an irreducible polynomial.

A2. No $f \in \mathfrak{P}$ is equal to a power $\geqslant 2$ of a polynomial.

Construct the code A (1)-(2) using $\mathfrak{P}$ as the defining set of polynomials. By Weil's inequality the sum of all coordinates of any $\mathbf{a} \in A$ differs from zero not more than by $(r-1)\sqrt{q}$. Add appropriate tails to all codewords of A to arrive to the balanced code B.

STATEMENT 1. *The code B has the parameters*

$$[k, n \leqslant q + (r-1)\sqrt{q}, M, d \geqslant \frac{k-1}{k}(q - 2r\sqrt{q}) - 2r], \quad S(B) < rp^{s/2}(1 + \log p)^s,$$

where $M = |\mathfrak{P}| \geqslant q^r(1 + o(1))$.

The main difficulty we meet whilst proving Lemma 2 is to estimate the number of polynomials that satisfy both A1 and A2. This is done following the ideas of ch.3 of [4].

3. Error correction and synchronization

Let us turn to the calculation of the code separation (4) of codes from Section 2. At first, define for a code $A[k, n, M, d]$ a parameter $\tau(A)$ which is closely connected with $\rho(A)$:

$$\tau(A) = \max_{\mathbf{a},\mathbf{b} \in A} \max_{1 \leqslant l \leqslant n-1} \tau_l(\mathbf{a}, \mathbf{b}),$$

where

$$\tau_l(\mathbf{a}, \mathbf{b}) = \sum_{i=0}^{n-l-1} \sum_{j=1}^{k} (a_{l+i}\bar{b}_i)^j.$$

Consider three codewords $\mathbf{a}, \mathbf{b}, \mathbf{c} \in A$. Then, clearly,

$$\operatorname{dist}(T_i(\mathbf{a}, \mathbf{b}), \mathbf{c}) = n - \frac{1}{k}(\tau_l(\mathbf{a}, \mathbf{c}) + \tau_{n-l}(\mathbf{c}, \mathbf{b})).$$

Therefore, an upper estimate of $\tau(A)$ provides a lower estimate for $\rho(A)$.

Let us show that certain subcodes of codes defined in Section 2 have large code separation. Consider the codes (2.1.3) with q a prime. To specify a subcode B' of a code B, form a set of polynomials $\mathfrak{P}' \subset \mathfrak{P}$, where $\mathfrak{P}$ is defined in 2.1.3. Of every q polynomials $f(x+c) \in \mathfrak{P}$, $c \in \mathbb{F}_q$, let us leave only one. Thus $|\mathfrak{P}'| = |\mathfrak{P}|/q$. For every f this corresponds to throwing out all $q-1$ cyclic shifts of the vector $\mathbf{a}_f$ (2). Now it is easy to see from (7) that the code $B' = A' \bigcup \overline{A'}$, where A' is constructed from $\mathfrak{P}'$ by (1)-(2), is comma-free with $\tau(B') \leqslant n + 2r\sqrt{q}(1+\log q)$ and

$$\rho(B') \geqslant \frac{n}{2} - 2r\sqrt{q}(1+\log q). \tag{9}$$

We summarize this argument in the following

STATEMENT. *Let q be an odd prime and $\mathfrak{P}'$ be the set of all monic square-free polynomials f over $\mathbb{F}_q$ with $1 \leqslant \deg f \leqslant r$ such that the inequality $f_1(x+\xi) \neq f_2(x)$, $\xi \neq 0$, holds for every pair $f_1, f_2 \in \mathfrak{P}$. Then the above-defined binary balanced comma-free code B' has length and distance* (6), *size $|B'| = 2q^{r-1}$, RDS* (8), *and code separation $\rho(B')$ given by* (9).

These results may be extended to a broader case. It is easy to see that for $q = p^s$, $s \geqslant 1$, a balanced dc-constrained code B, defined in 2.1.3, is comma-free with $\rho(B) \geqslant n/2 - 2rp^{s/2}(1+\log p)^s$. A similar estimate holds also for k-ary codes from 2.2.2.

4. CONCLUSION

We have systematically investigated codes with vectors formed by values of a multiplicative character of a finite field, accepted on values of a polynomial with argument running over the points of this field. The k-ary codes that are constructed, have rate $R \to 0$ for $n \to \infty$, distance $d \gtrsim (k-1)n/k - \alpha_1(n)\sqrt{n}$, where $\alpha_1(n)$ is a slowly growing function of n. Besides, these codes are balanced dc-constrained with RDS$\cong \alpha_2(n)\sqrt{n}$ and comma-free (except for the 'linear' codes such as Hadamard and the like) with the code separation $\rho(\cdot) \gtrsim n/2 - \alpha_3(n)\sqrt{n}$. Here α_2 and α_3 also slowly grow with n.

References

[1] R.P.Bambah,D.D.Joshi, and I.S.Luthar. Some lower bounds on the number of code points in a minimum distance binary codes I,II. *Inform. and Contr.*, **4**,4 (December 1961), 313–323.

[2] L.A.Bassalygo. On separable comma-free codes. *Problemy Peredachi Inform.*, **2**,4 (1966), 78–79, and *Probl. Inform. Trans.* **2** (1966), 52–53.

[3] A.M.Barg and S.N.Litsyn DC-constrained codes from Hadamard matrices. *IEEE Trans. Inform. Theory*, **37**,3,Pt.2 (May 1991), 801–807.

[4] E.R.Berlecamp. *Algebraic Coding Theory.* McGraw Hill, N.Y. et al., 1968.

[5] S.Golomb, W.Gordon, and L.Welch. Comma-free codes. *Canad. J. Math.*, **10**,2 (1958), 202–209.

[6] V.I.Levenstein. Bounds for codes that provide error correction and synchronization. *Problemy Peredachi Inform.*,**5**,2 (1969), 3–13, and Probl. Inform. Trans. **5** (1969).

[7] V.I.Levenstein. One method of constructing quasilinear codes providing synchronization in the presence of errors. *Problemy Peredachi Inform.*, **7**,3 (1971), 30–40, and *Probl. Inform. Trans*, **7** (1971), 215–227.

[8] F.J.MacWilliams and N.J.A.Sloane. *The Theory of Error-Correcting Codes.* North-Holland, Amsterdam, 1977.

[9] V.M.Sidel'nikov. Some k-valued pseudo-random sequences and nearly equidistant codes. *Problemy Peredachi Inform.*, **5**,1 (1967),16–22, and *Probl. Inform. Trans.*, **5** (1969), 12–16.

[10] I.M.Vinogradov. *Elements of Number Theory.* 9th ed., Moscow, 1981, in Russian.

CONSTRUCTIONS OF CODES WITH COVERING RADIUS 2

Alexander A. Davydov

Institute for Problems of Cybernetics of the USSR Academy of Sciences, Vavilov street 37, Moscow 117312, USSR

Abstract. Constructions of nonbinary linear codes with covering radius $R = 2$ are considered. Infinite families of linear q-ary codes with $R = 2$, $q \geq 4$, and a table of quaternary linear codes with $R = 2$, redundancy $r \leq 20$, are given.

1. INTRODUCTION.

We consider linear covering codes over the Galois field $GF(q)$, $q \geq 2$. Linear covering codes are being extensively studied, see, e.g., [1]-[7]. (See also references in [1]-[7]).

In this paper developing and using results of the works [1], [4]-[6], we consider constructions of q-ary linear codes of covering radius $R = 2$. These constructions based upon a starting code V_0 with $R = 2$ form an infinite family of codes of the same covering radius. We obtain families of linear codes with $R = 2$, $q \geq 4$, and a table of codes with $R = 2$, $q = 4$, redundancy $r \leq 20$. Constructed codes have better parameters than a direct sum [7] of Hamming codes.

Denote by an $[n,n-r]_qR$ code a q-ary linear code of length n, redundancy r, covering radius R, and cardinality q^{n-r}. Let E_q^n be the space of n-dimensional vectors over the field $GF(q)$. For an $[n,n-r(C)]_qR$ code C denote by $\mu_q(n,R,C)$ the density of a covering of the space E_q^n by spheres of radius R, whose centers are codewords of C [3]-[6].

$$\mu_q(n,R,C) \triangleq q^{n-r(C)} \sum_{i=0}^{R} (q-1)^i \begin{bmatrix} n \\ i \end{bmatrix} / q^n. \tag{1.1}$$

Let U be an infinite family of q-ary codes with covering radius R, and let U_n be a code of length n, $U_n \in U$. For the family U we consider the following value [4]-[6] :

$$\bar{\mu}_q(R,U) \triangleq \liminf_{n \to \infty,\, U_n \in U} \mu_q(n,R,U_n) . \tag{1.2}$$

The direct sum of the $[\mathcal{f}_{t,q}, \mathcal{f}_{t,q} - t]_q 1$ Hamming codes, where

$$\mathcal{f}_{t,q} \triangleq (q^t - 1) / (q - 1) , \tag{1.3}$$

gives a family Ω consisting of $[n_r^*, n_r^* - r]_q 2$ codes HN_r^* with

$$n_r^* = 2\mathcal{f}_{t,q} \quad \text{if } r = 2t, \qquad n_r^* = \mathcal{f}_{t,q} + \mathcal{f}_{t+1,q} \quad \text{if } r = 2t+1. \tag{1.4}$$

For the family Ω of codes HN_r^* we have $\bar{\mu}_q(2,\Omega) \approx 2$. So, it is interesting to obtain infinite families of codes with $\bar{\mu}_q(2,U) < 2$.

In [6] linear codes having $\bar{\mu}_2(2,U) = 49 / 32$ are constructed. In [4] families with $\bar{\mu}_2(2,U) \approx 1.477$, $\bar{\mu}_3(2,U) \approx 1.185$ are obtained.

In this paper we construct an infinite family A_q with $\bar{\mu}_q(2,A_q) < 2$ for $q \geq 4$. The family A_q consists of $[n,n-r]_q R$ codes having the following parameters:

$$R = 2, \quad q \geq 4, \quad r = 2t ; \qquad \bar{\mu}_4(2,A_4) \approx 1.64, \quad \bar{\mu}_5(2,A_5) \approx 1.69;$$

$$\bar{\mu}_q(2,A_q) \approx 2 - 2q^{-1} + 2.5q^{-2} - 0.5q^{-3} - 1.5q^{-4} \ldots < 2 ,$$

$$n = 2\mathcal{f}_{t,q} - 1 \quad \text{for } t = 2,3; \qquad n = 2\mathcal{f}_{t,q} - q^{t-2} \quad \text{for } t = 4,5;$$

$$n = 2\mathcal{f}_{t,q} - q^{t-2} - \mathcal{f}_{t-5,q} \quad \text{for } t = 8;$$

$$n = 2\mathcal{f}_{t,q} - q^{t-2} - \mathcal{f}_{t-4,q} \quad \text{for } t = 6,7 \text{ and } t \geq 9. \tag{1.5}$$

The length function $l(r,R;q)$ is the smallest length of a q-ary linear code with redundancy r and covering radius R [1]. The obtained codes give upper bounds on $l(r,2;q)$.

2. CONSTRUCTIONS OF CODES

All columns and matrices are q-ary. An element h of $GF(q^m)$ written as an element of a q-ary matrix denotes the column m-dimensional vector that is the q-ary representation of h. An upper index in a denotation of a matrix (resp. column) is the number of rows (resp. components) in it.

Let $P^s(\varphi,q^m) \triangleq [\varphi \ \dots \ \varphi]$ be a $s \times q^m$ matrix of equal columns φ. Denote by $0^s(u)$ the zero $s \times u$ matrix. Let W_q^m be the parity check matrix of the $[f_{m,q}, f_{m,q} - m]_q 1$ Hamming code. Let

$$B^{2m}(b;q) \triangleq \begin{bmatrix} e_0 & e_1 & \dots & e_N \\ e_0 b & e_1 b & \dots & e_N b \end{bmatrix}, \qquad B^{2m}(*;q) \triangleq \begin{bmatrix} & 0^m(N) & \\ \hline e_0 & e_1 & \dots & e_N \end{bmatrix},$$

where $N = q^m - 1$; $e_i, b \in GF(q^m)$, $i = \overline{0,N}$; $e_i \neq e_j$ if $i \neq j$.

$$D_1^{2m} \triangleq \begin{bmatrix} 0^m(f_{m,q}) \\ W_q^m \end{bmatrix}; \qquad D_2^{2m} \triangleq \begin{bmatrix} W_q^m & 0^m(f_{m,q}) \\ 0^m(f_{m,q}) & W_q^m \end{bmatrix}.$$

Fact 2.1 [2]: A linear q-ary code of redundancy r, with a parity check matrix H^r has covering radius 2 if any q-ary column of length r is a linear combination (with coefficients from $GF(q)$) of two or fewer columns of the matrix H^r.

The proposed constructions form an infinite family of *new codes* V on the base of a *starting code* V_0.

A *starting code* V_0 is an $[Y, Y - s]_q 2$ code with a parity check matrix

$$\Phi_0^s = [\varphi_1 \ \varphi_2 \ \dots \ \varphi_Y]. \tag{2.1}$$

Denote by J_0 a set of linear combinations of two or fewer columns of the matrix Φ_0^s such that any q-ary column of length s is equal to some linear combination of this set. We construct a graph $\Gamma(J_0)$ associated to the set J_0. The graph $\Gamma(J_0)$ has Y vertices. The vertex with the number t corresponds to the column φ_t of the matrix Φ_0^s. Vertices t and k are connected by an edge if and only if both columns φ_t and φ_k simultaneously belong to some linear combination of the set J_0.

Denote by $h(J_0)$ the chromatic number of graph $\Gamma(J_0)$, i.e., the minimum number of colors necessary to color vertices, with connected vertices colored differently. $h(J_0) \le Y$.

A *new code* V is a code with the parity check matrix

$$H_\gamma^{s+2m}(J_0) = \left[\begin{array}{c|ccc} 0^s(\gamma f_{m,q}) & P^s(\varphi_1, q^m) & \dots & P^s(\varphi_Y, q^m) \\ D_\gamma^{2m} & B^{2m}(b_1;q) & \dots & B^{2m}(b_Y;q) \end{array}\right], \tag{2.2}$$

where $\gamma \in \{1,2\}$; if the vertices of $\Gamma(J_0)$ with numbers t and k are connected by an edge then the relation $b_t \ne b_k$ holds; if vertices of $\Gamma(J_0)$ with numbers u and v are *not* connected by an edge then we are free to assign the equality $b_u = b_v$ or the inequality $b_u \ne b_v$;

$$\text{if } \gamma = 1 \text{ then } Y \ge q^m \ge h(J_0), \ \{b_1 \cup b_2 \cup \dots \cup b_Y\} = GF(q^m) \ ;$$

$$\text{if } \gamma = 2 \text{ then } q^m + 1 \ge h(J_0), \ \{b_1 \cup \dots \cup b_Y\} \subseteq \{GF(q^m) \cup *\}.$$

The parameter m is bounded from below, but it *not bounded from above*, i.e., we obtain an *infinite family* of codes V.

Theorem 2.1: The new code V with parity check matrix $H_\gamma^{s+2m}(J_0)$ is an $[n,n-r]_q R$ code with parameters

$$R = 2, \qquad n = Yq^m + \gamma f_{m,q}, \qquad r = s + 2m, \qquad \gamma \in \{1,2\}. \tag{2.3}$$

Proof: Values of n and r follow from (2.2). Relations $q^m \geq h(J_0)$ and $q^m + 1 \geq h(J_0)$ provide an assignment $b_i \neq b_k$ if vertices of $\Gamma(J_0)$ with numbers i and k are connected.

We show that $R = 2$ applying Fact 2.1.

Let U^{s+2m} be arbitrary q-ary column of length $s + 2m$,

$$U^{s+2m} = \begin{bmatrix} v \\ u \\ l \end{bmatrix}, \qquad v^s \in GF(q^s), \qquad u,l \in GF(q^m). \tag{2.4}$$

The following cases should be considered: $v = c\varphi_i + d\varphi_k$, $v = g\varphi_j$, $v = 0$, where $c,d,g \in GF(q)$, the linear combination $c\varphi_i + d\varphi_k$ belongs to set J_0. It is easy to see that in all cases the column U^{s+2m} equals a linear combination of two or fewer columns of matrix $H_\gamma^{s+2m}(J_0)$. For example, when $\gamma = 1$ we have for syndromes of vectors at distance 2 from V :

$$\begin{bmatrix} v \\ u \\ l \end{bmatrix} = c\begin{bmatrix} \varphi_i \\ x \\ xb_i \end{bmatrix} + d\begin{bmatrix} \varphi_k \\ y \\ yb_k \end{bmatrix}, \qquad \begin{bmatrix} v \\ u \\ l \end{bmatrix} = g\begin{bmatrix} \varphi_j \\ ug^{-1} \\ ug^{-1}b_j \end{bmatrix} + a\begin{bmatrix} 0^s \\ 0^m \\ z \end{bmatrix},$$

$$\begin{bmatrix} v \\ u \\ l \end{bmatrix} = \begin{bmatrix} \varphi_t \\ u \\ ub_t = l \end{bmatrix} - \begin{bmatrix} \varphi_t \\ 0^m \\ 0^m \end{bmatrix}, \qquad x,y,z \in GF(q^m),\ a \in GF(q), \tag{2.5}$$

where 0^e is the zero column of length e.

The values of x,y,z,a, and b_t can be found since $b_i \neq b_k$, the Hamming code has $R = 1$, and $\{b_1 \cup b_2 \cup \ldots \cup b_Y\} = GF(q^m)$. □

3. THE INFINITE FAMILY A_q WITH $\bar{\mu}_q(2,A_q) < 2$, $q \geq 4$

We generalize for odd q the code obtained in [1, p.104] for even q.

Theorem 3.1: Suppose that an $[n,n-r]_q R$ code Φ_0 with $q \geq 4$ has the following parity check matrix

$$H^4 = \left[\begin{array}{ccccc|c|ccccc} 1 & 1 & 1 & \dots & 1 & 0 & 0 & 0 & 0 & \dots & 0 \\ 0 & 1 & a_2 & \dots & a_{q-1} & 1 & 0 & 0 & 0 & \dots & 0 \\ 0 & 1 & a_2^2 & \dots & a_{q-1}^2 & 0 & 0 & 1 & 1 & \dots & 1 \\ 0 & 0 & 0 & \dots & 0 & 0 & 1 & 1 & a_2 & \dots & a_{q-1} \end{array}\right], \qquad (3.1)$$

where $0,1,a_i \in GF(q)$, $a_i \neq 0$, $a_i \neq 1$, $i = \overline{2,q-1}$, $a_i \neq a_j$ if $i \neq j$. Then the code Φ_0 has parameters

$$R = 2, \qquad q \geq 4, \qquad r = 4, \qquad n = 2q + 1 = 2f_{2,q} - 1. \qquad (3.2)$$

Proof: For q even see [1, p.104]. For q odd we can directly use Fact 2.1. Let $a,b,c,d \neq 0$, $a,b,c,d \in GF(q)$. Then, e.g.,

$$\begin{bmatrix} 0 \\ a \\ b \\ 0 \end{bmatrix} = \begin{bmatrix} 1 \\ x \\ x^2 \\ 0 \end{bmatrix} - \begin{bmatrix} 1 \\ y \\ y^2 \\ 0 \end{bmatrix}, \qquad \begin{bmatrix} a \\ b \\ c \\ d \end{bmatrix} = a \begin{bmatrix} 1 \\ z \\ z^2 \\ 0 \end{bmatrix} + t \begin{bmatrix} 0 \\ 0 \\ 1 \\ u \end{bmatrix},$$

where $x = (b + a^2)/2a$, $y = x - a$, $z = a^{-1}b$, $t = c - az^2$, $u = t^{-1}d$. □

We construct the family A_q with parameters (1.5) iteratively using obtained matrices $H_\gamma^{s+2m}(J_0)$ as matrices Φ_0^{s+2m} for the next step of the iterative process. If in (2.1) the graph $\Gamma(J_0)$ for the matrix Φ_0^s has chromatic number $h(J_0)$ then for the matrix $H_\gamma^{s+2m}(J_0)$ treated as a new matrix Φ_0^{s+2m} we can obtain a graph $\Gamma(J_0')$ with chromatic number $h(J_0')$ such that

$$h(J_0') \le 2q^m + 1 \quad \text{if} \quad \gamma = 1, \quad h(J_0') \le h(J_0) + 2 \quad \text{if} \quad \gamma = 2. \quad (3.3)$$

One can prove relations (3.3) taking into account the structures of the matrices $H_\gamma^{s+2m}(J_0)$ and techniques of representation of column U^{s+2m} by a linear combination of two columns (see (2.5)).

On the first step we use code Ψ_0 of (3.2) as V_0 with $s = 4$, $\Phi_0^s = H^4$, $h(J_0) = Y = 2q + 1$. Take $\gamma = 2$, $m = 2,3$. We obtain codes $\Psi_{1,m}$ of length $Y_{1,m} = (2q + 1)q^m + 2f_{m,q} = 2f_{m+2,q} - q^m$ (cf. (1.5) for $t = 4,5$), with $h(J_0') = 2q + 3$ (see (3.3)).

On the second step we use $\Psi_{1,2}$ as code V_0. Inequalities $Y_{1,2} \ge q^m \ge h(J_0')$ hold for $m = 2,3$. Hence we can take $\gamma = 1$, $m = 2,3$. We obtain codes $\Psi_{2,m}$ of length $Y_{2,m} = Y_{1,2}q^m + f_{m,q} = 2f_{m+4,q} - q^{m+2} - f_{m,q}$ (cf. (1.5) for $t = 6,7$), with $h(J_0') = (2q + 3)q^m + 1$ (see (3.3)).

On the third step we take codes $\Psi_{2,m}$ as V_0, $\gamma = 1$, etc.

Then we form an iterative infinite chain of codes taking $\gamma = 1$ and checking relations $Y_{t,m} \ge q^m \ge h(J_0')$.

For $t = 8$ in (1.5) on the 2nd step we take code $\Psi_{1,3}$ as V_0.

Finally, for $t = 3$ in (1.5) we proceed in the following way.

Let the starting code V_0 be a code HN_s^* of (1.4), $s \ge 3$. It is easy to see that $h(J_0) = 2$ and $Y = n_s^* > q^1 \ge h(J_0)$. Using the construction (2.2) with $\gamma = 1$, $m = 1$, we obtain a $[n, n - r]_q R$ code having (for even and odd r) parameters

$$R = 2, \quad q \ge 4, \quad r \ge 5, \quad n = n_r^* - 1, \quad h(J_0) = 2q + 1. \quad (3.4)$$

4. QUATERNARY CODES WITH $R = 2$

From (3.4) we obtain a $[25,25-5]_4 2$ code. Using it as code V_0 and forming an iterative code chain by the construction (2.2) with $\gamma = 1$ (see Section 3) we produce an infinite family B_4 consisting of $[n,n-r]_q R$ codes with parameters

$$R = 2, \quad q = 4, \quad r = 2t - 1, \quad t = 3,5,8,9, \quad \text{and } t \geq 12,$$

$$n = (76 \times 4^{t-3} - 1) / 3. \tag{4.1}$$

On the base of relations (1.5),(2.3),(3.2),(3.4), and (4.1) we obtain Table I. All entries in the table for even r follow from (1.5). It is known that a $[6,6-3]_4 2$ code exists. Codes with $r = 5,7$ are obtained from (3.4). Entries for $r = 9,15,17$ follow from (4.1). Codes having $r = 11,13$ are designed by the construction (2.2) with $\gamma = 2$, $m = 3,4$. A $[25,25-5]_4 2$ code is used as V_0. For $r = 19$ we take in (2.2) $\gamma = 1$, $m = 4$, and the $[1642,1642-11]_4 2$ code from Table I as code V_0.

TABLE I

The Least Known Length n of Linear Quaternary $[n,n-r]_4 2$

Codes with Redundancy r and Covering Radius 2

r	n	r	n	r	n	r	n
		6	41	11	1642	16	39573
2	2	7	105	12	2469	17	103765
3	6	8	154	13	6570	18	158037
4	9	9	405	14	9877	19	420437
5	25	10	618	15	25941	20	632149

REFERENCES

[1] R. A. Brualdi, V. S. Pless, and R. M. Wilson, "Short codes with a given covering radius," *IEEE Trans. Inform. Theory*, vol. IT-35, no. 1, pp. 99-109, Jan. 1989.

[2] G. D. Cohen, M. G. Karpovsky, H. F. Mattson,Jr.,and J. R. Shatz, "Covering radius - Survey and recent results," *IEEE Trans. Inform. Theory*, vol. IT-31, no. 3, pp. 328-343, May 1985.

[3] G. D. Cohen, A. C. Lobstein, and N. J. A. Sloane, "Further results on the covering radius of codes," *IEEE Trans. Inform. Theory*, vol. IT-32, no. 5, pp. 680-694, Sept. 1986.

[4] A. A. Davydov,"Construction of linear covering codes," *Probl. Peredach. Inform.*, vol. 26, no. 4, pp. 38-55, Oct.-Dec.1990. (In Russian).

[5] A. A. Davydov,"Constructions and families of q-ary linear covering codes and saturated sets of points in projective geometry," in *Proc.* V *Joint Soviet-Swedish Intern. Workshop on Inform. Theory " Convolutional Codes; Multi-user Communication"*, Moscow, USSR, pp. 46-49, Jan. 1991.

[6] E. M. Gabidulin, A. A. Davydov, and L. M. Tombak, "Codes with covering radius 2 and other new covering codes," *IEEE Trans. Inform. Theory*, vol. IT-37, no. 1, pp. 219-224, Jan. 1991.

[7] R. L. Graham and N. J. A. Sloane, "On the covering radius of codes," *IEEE Trans. Inform. Theory*, vol. IT-31, no. 3, pp. 385-401, May 1985.

ON PERFECT WEIGHTED COVERINGS WITH SMALL RADIUS

Gérard D. Cohen
Ecole Nationale Supérieure des Télécommunications
46 rue Barrault, C-220-5
75634 Paris cédex 13, France
Email: Cohen@inf.enst.fr

Simon N. Litsyn
Dept. of Electrical Engineering-Systems
Tel-Aviv University
Ramat-Aviv
69978, Israel
Email: litsyn@genius.tau.ac.il

H. F. Mattson, Jr.
School of Computer and Information Science
4-116 Center for Science & Technology
Syracuse, New York 13244-4100
Email: jen@SUVM.acs.syr.edu, jen@SUVM.bitnet

Abstract: We extend the results of our previous paper [8] to the nonlinear case: The Lloyd polynomial of the covering has at least R distinct roots among $1, \ldots, n$, where R is the covering radius. We investigate *PWC* with diameter 1, finding a partial characterization. We complete an investigation begun in [8] on linear *PMC* with distance 1 and diameter 2.

1 Introduction

Much attention has been devoted to the problem of classifying perfect codes (See [13, 15]). Further generalizations of perfectness were introduced in [10, 2, 11, 14]. For all these codes the diameter of the covering spheres equals the covering radius of the code which by use of Delsarte's results leads to a very rigid set of possible parameters. This framework was broadened by introducing new types of perfect configurations [5, 6, 12, 16]. All these extensions fall under the concept of perfect weighted coverings (*PWC*) first considered in [8]. Although general, these definitions leave hope for a complete classification, at least for small diameter. The linear case with diameter at most 2 was considered in [8], where some motivation related to list decoding was given.

We are pleased to acknowledge that this problem arose in discussions with I. Honkala in Veldhoven in June, 1990.

2 Notations and known results

We denote by $\boldsymbol{F}^n$ the vector space of binary n-tuples, by $d(\cdot,\cdot)$ the Hamming distance, by $C(n,K,d)R$ a code C with length n, size K, minimum distance $d = d(C)$ and covering radius R [9], [7]. When C is linear, we write $C[n,k,d]R$, where k is the binary log of K. We denote the Hamming weight of $x \in \boldsymbol{F}^n$ by $|x|$.

For $x \in \boldsymbol{F}^n$, $A(x) = (A_0(x), A_1(x) \dots A_n(x))$ will stand for the distance distribution of C with respect to x; thus

$$A_i(x) := |\{c \in C : d(c,x) = i\}|.$$

For any $(n+1)$-tuple $M = (m_0, m_1, \dots, m_n)$ of weights, i.e., rational numbers, we define the *M-density* of C at x as

$$\theta(x) := \sum_{i=0}^{n} m_i \, A_i(x) = \langle M, A(x) \rangle. \tag{2.1}$$

We consider only *coverings*, i.e., codes C such that $\theta(x) \geq 1$ for all x.

(2.2) C is a *perfect M-covering* if $\theta(x) = 1$ for all x.

We define the *diameter* of an M-covering as

$$\delta := \max\{i : m_i \neq 0\}.$$

To avoid trivial cases, we usually assume that $m_i = 0$ for $i \geq n/2$, i.e., $\delta < n/2$.

Here are the known special cases.

(2.3) Classical perfect codes: $m_i = 1$ for $i = 0, 1, \dots \delta$.

(2.4) Perfect multiple coverings (*PMC*): $m_i = 1/j$ for $i = 0, 1, \dots \delta$ where j is a positive integer. See [16] and [5].

(2.5) Perfect L-codes: $m_i = 1$ for $i \in L \subseteq \{0, 1, \dots \lfloor n/2 \rfloor\}$. See [12] and [6].

(2.6) Strongly uniformly packed codes:
$m_i = 1$ for $i = 0, 1, \dots, e-1$
$m_e = m_{e+1} = 1/r$ for some integer r. See [14].

(2.7) Uniformly packed codes [2, 11]. For these codes $\delta(M) = R(C)$, and the m_i's are uniquely determined.

The following necessary and sufficient condition was already in [8] in the linear case. For a perfect M-covering C one gets from the definition:

$$\sum_{i=0}^{n} m_i \; A_i(x) = 1 \text{ for all } x.$$

Summing over all x in $\boldsymbol{F}^n$ and permuting sums, we get

$$\sum_{i=0}^{n} m_i \sum_{x \in F^n} A_i(x) = 2^n.$$

For $i = 0$, the second sum is $|C| = K$, for $i = 1$ it is Kn, and so on. For the converse we use the condition $\theta(x) \geq 1$. Hence we get the following analog of the Hamming condition.

Proposition 2.1 *A covering C is a perfect M-covering if and only if*

$$K \sum_{i=0}^{n} m_i \binom{n}{i} = 2^n . \tag{2.8}$$

□

3 A Lloyd theorem

In this section we prove

Theorem 3.1 *Let C be a perfect weighted covering with $M = (m_0, m_1, \dots, m_\delta)$. Then the Lloyd polynomial of this covering,*

$$L(x) := \sum_{0 \leq i \leq \delta} m_i \; P_{n,i}(x)$$

has at least R distinct integral roots among $1, 2, \dots, n$.

Proof. (Adapted from [1], Chapter II, Section 1, which records A. M. Gleason's proof of the classical Lloyd theorem.) The first part of the proof is identical to that of [8, Thm. 4.1].

We use the group algebra $\mathcal{A}$ of all formal polynomials

$$\sum_{a \in \boldsymbol{F}^n} \gamma_a X^a$$

with $\gamma_a \in \boldsymbol{Q}$, the field of rational numbers.

Define

$$S := \sum_{0 \le i \le \delta} m_i \sum_{|a|=i} X^a. \tag{3.1}$$

We let the symbol C for our code also stand for the corresponding element in $\mathcal{A}$, namely,

$$C := \sum_{c \in C} X^c. \tag{3.2}$$

Then we find that

$$SC = \sum_{c \in C} X^c \cdot S \ = \ \boldsymbol{F}^n := \sum_{a \in \boldsymbol{F}^n} X^a. \tag{3.3}$$

Characters on $\boldsymbol{F}^n$ are group homomorphisms of $(\boldsymbol{F}^n, +)$ into $\{1, -1\}$, the group of order 2 in $\boldsymbol{Q}^\times$. All characters have the form χ_u for $u \in \boldsymbol{F}^n$, where χ_u is defined as

$$\chi_u(v) \ = \ (-1)^{u \cdot v} \ \text{ for } \ u, v \in \boldsymbol{F}^n.$$

We use linearity to extend χ_u to a linear functional defined on $\mathcal{A}$:
For all $Y \in \mathcal{A}$ if $Y = \sum_{a \in \boldsymbol{F}^n} \gamma_a X^a$, then $\chi_u(Y) := \sum \gamma_a \chi_u(a)$.
It follows that

$$\chi_u(YZ) \ = \ \chi_u(Y)\chi_u(Z) \ \text{ for all } Y, Z \in \mathcal{A}.$$

It is known [1, 9] that for any $u \in \boldsymbol{F}^n$, if $|u| = w$, then

$$\chi_u\left(\sum_{|a|=i} X^a\right) \ = \ P_{n,i}(w). \tag{3.4}$$

It follows that

$$\chi_u(S) \ = \ L(w). \tag{3.5}$$

¿From (3.3), furthermore, we see that

$$\chi_u(SC) \ = \ \chi_u(S)\chi_u(C) \ = \ 0$$

for all $u \neq 0$.

Let $u_0, u_1, \ldots, u_R$ be translate-leaders for C such that $|u_i| = i$. Define

$$C_i := X^{u_i}\, C.$$

Then

$$S\, C_i \ = \ \mathbf{F}^n. \tag{3.6}$$

Define the symmetric subring $\overline{\mathcal{A}}$ of $\mathcal{A}$ as the set of all elements Y of $\mathcal{A}$ in which the coefficient of X^a depends only on the weight of a:

$$Y \ = \ \sum_{a \in \mathbf{F}^n} \gamma_a\, X^a \ \in \overline{\mathcal{A}} \ \text{ iff } \ \forall a, b \in \mathbf{F}^n, \ |a| = |b| \rightarrow \gamma_a = \gamma_b. \tag{3.7}$$

The mapping $T : \mathcal{A} \to \overline{\mathcal{A}}$ defined by

$$T(Y) := \frac{1}{n!} \sum_{\varphi} \varphi(Y),$$

where φ runs over all $n!$ permutations of the n coordinates of F^n, maps $\overline{\mathcal{A}}$ onto $\overline{\mathcal{A}}$. Furthermore, as the reader may easily verify,

$$\forall Y \in \overline{\mathcal{A}},\ \forall Z \in \mathcal{A},\ T(YZ) = YT(Z). \tag{3.8}$$

Define $\overline{C}_i := T(C_i)$. Applying (3.8) to (3.6), we see that

$$S\overline{C}_i = \mathbf{F}^n$$

since, of course, $S \in \overline{\mathcal{A}}$. Define also

$$K := \{Z;\ Z \in \overline{\mathcal{A}},\ SZ = 0\}. \tag{3.9}$$

Thus K is the kernel of the linear mapping from $\overline{\mathcal{A}}$ to $\overline{\mathcal{A}}$ defined by $Y \longmapsto SY$ for all $Y \in \overline{\mathcal{A}}$.

It follows from (3.8) that for any character χ_u such that $\chi_u(S) \neq 0$,

$$\forall Z \in K,\ \chi_u(Z) = 0.$$

Since $\overline{\mathcal{A}}$ has dimension $n+1$, its space of linear functionals also has dimension $n+1$. Since every linear functional on $\overline{\mathcal{A}}$ can be extended to one on $\mathcal{A}$, the $n+1$ linear functionals on $\overline{\mathcal{A}}$ obtained by restricting the χ_u to $\overline{\mathcal{A}}$, as

$$\chi_{u|_{\overline{\mathcal{A}}}} =: \chi_w \text{ for } |u| = w$$

$$w = 0, 1, \ldots, n,$$

are linearly independent.

Suppose that ρ is the exact number of values of $w \in \{0, 1, \ldots, n\}$ for which

$$\chi_w(S) \neq 0.$$

Since $\chi_w(S)\chi_w(K) = 0$ for all w, it follows that $\chi_w(K) = 0$ for ρ values of w. Since $S\overline{C}_i = \mathbf{F}^n$ for $i = 0, 1, \ldots, R$, we see that

$$S(\overline{C}_i - \overline{C}_0) = 0 \text{ for } i = 1, \ldots, R.$$

The elements $\overline{C}_i - \overline{C}_0$ are linearly independent because $\overline{C}_i$ contains elements of weight i but of no smaller weight. We find that

$$R \leq \dim_{\mathbf{Q}} K \leq n + 1 - \rho,$$

since K is included in the intersection of the t kernels of the χ_w mentioned above. But $n+1-\rho$ is the number of χ_w's which vanish on S; therefore $\chi_w(S) = 0$ for at least R values of w.

Notice now that

$$\chi_w(S) = \sum_{0 \leq i \leq \delta} m_i\, P_{n,i}(w).$$

This finishes the proof. □

4 A construction

Definition 4.1 *Let $C(n,K,d)R$ and $C'(n',K',d')R'$ be two codes. Set*

$$\chi_C(x) = \begin{cases} 0 & \text{if } x \in C \\ 1 & \text{otherwise.} \end{cases}$$

We extend χ_C to a mapping $\chi : \mathbf{F}^{nn'} \to \mathbf{F}^{n'}$ by setting

$$\chi(x) := (\chi_C(x_1), \chi_C(x_2), \ldots \chi_C(x_{n'}))$$

where the x_i's are in $\mathbf{F}^n$, for $1 \le i \le n'$, and $x = (x_1, x_2, \ldots x_{n'})$ is their concatenation. We are now ready to define $C \otimes C'$ as follows:

$$C \otimes C' = \left\{ z \in \mathbf{F}^{nn'} : \chi(z) \in C' \right\}.$$

Proposition 4.1 *$C \otimes C'$ has length nn', minimum distance $\min\{d, d'\}$ and covering radius RR'.*

The proof is immediate. □

Proposition 4.2 *Let x and x' be such that $d(x,C) = R, d(x',C') = R'$. Suppose that $A_R(x)$ and $A'_{R'}(x')$ are independent of x. Then for $C \otimes C'$ the coefficient $A_{RR'}(z)$ is the same for any z such that $d(z, C \otimes C') = RR'$ and one has*

$$A_{RR'} = A_R A'_{R'}.$$

□

5 *PWC* with diameter 1

Let us denote such a *PWC* by (n, m_0, m_1). From (2.2), $A_1(x) = 1/m_1$ for any x not in C. Hence $m_1 = 1/p$, where p is an integer. This means that every two noncodewords have the same number of codewords at distance 1.

For $c \in C$, we get: $m_0 + A_1(c)/p = 1$, hence

$$A_1(c) = p\ (1 - m_0)$$

is a constant independent of c. Since $A_1(c)$ is an integer , so is $m_0 p$.

Now the Hamming analogue (2.9) gives

$$K\ (pm_0 + n) = p\ 2^n,$$

which implies

(5.1) $$n = p'\ 2^i - m_0 p, \text{ with } p' \mid p.$$

The case $m_0 = 1/p$ corresponds to the *PMC* mentioned in (2.4); it is solved in [16] and [8].

Let us give a few general constructions.

Proposition 5.1 *If there exists a PWC* $C(n, m_0, m_1)$, *then for any* $l \geq 0$ *there exists a PWC* $C'(n+l, m_0 - lm_1, m_1)$.

Proof. Let us define C' as the set of vectors (c, f) in F^{n+l}, where $c \in C$ and $f \in F^l$. Let A be the distance distribution for C_1 and A' that for C'. There are two possibilities for an arbitrary $(x, f) \in F^{n+l}$:

(a) $x \in C$. Then $A'_1((x, f)) = A_1(x) + l$. Evidently $A'_0((x, f)) = 1$.

(b) $x \notin C$. Then $A'_0((x, f)) = 0$ by construction and $A'_1((x, f)) = A_1(x)$. □

Proposition 5.2 *If there exists a PWC* $C(n, m_0, m_1)$, *then there exists a PWC* $C'(ns, m_0, m_1/s)$.

Proof. Apply construction ⊗ (Def. 4.1) with outer code $C(n, m_0, m_1)$ and inner code the $[s, s-1]$ parity code. □

Proposition 5.3 *If there exists a PWC* $C(n, m_0, m_1)$, *then there exists a PWC* $C'(n, m_0/i, m_1/i)$, *for* i *a positive integer.*

Proof. Take the union of i cyclic shifts of code C. □

Let us now turn to the special case when $m_0 = 1$.

Proposition 5.4 *A PWC with* $\delta = m_0 = 1$ *exists for* $n = p(2^i - 1), m_1 = 1/p$. *It can be achieved by a linear code.* □

See [8] for a proof of this result. In contrast to the linear case, [8, Prop. 5.4], we cannot characterize *PWC* with $\delta = m_0 = 1$ here. However, we have a partial characterization:

Proposition 5.5 *A PWC* $(n, 1, 2^{-q})$ *exists if and only if for some* i $n = 2^q(2^i - 1)$. *Such a PWC can be achieved by a linear code.*

Proof. If $m_1 = 2^{-q}$, then $p' = 2^{q'}, q' \leq q$, and (5.1) gives $n = 2^q(2^{i+q'-q} - 1)$. The converse stems from Proposition 5.4. □

We would like to point out that for some parameters satisfying (5.1) there is no corresponding code.

Consider the case $m_0 = 1$, $m_1 = 1/3$. Proposition 5.4 gives the sequence of lengths $n = 3 \cdot 2^i - 3$. The other possibility is $n = 2^i - 3$. The first code in this sequence would be a *PWC* with $n = 5$ and $K = 12$. Let us show its nonexistence.

Proposition 5.6 *A* $(5, 1, 1/3)$ *PWC does not exist.*

Proof. We may assume the code contains the zero vector. Furthermore, it does not contain vectors of weight 1, since the minimum distance is 2 for $m_0 = 1$. Every vector of weight 1 has to be covered by exactly two codewords of weight 2. There are exactly 5 codewords of weight 2, because if we consider the matrix of all such codewords, we see that each column has sum 2 (by the "coverage" condition just mentioned). Let x be any vector in $\mathbf{F}^5$ of weight 3. Each "1" in x is covered by two codewords of weight 2. That makes six codewords of weight 2. By the pigeonhole principle, two are equal, say to $c \in C$. Then x is at distance 1 from c.

So the code does not contain vectors of weight 1 and 3, and we cannot cover vectors of weight 2. □

6 Linear *PMC* with diameter 2 ($m_0 = m_1 = m_2 = 1/j$)

The purpose of this section is to summarize and extend results from [8].

6.1 The case $s = 1$

Proposition 6.1 [8] *The only PMC with* $s = 1, d = 2$ *is the* $[2,1,2]$ *code with* $j = 2$. □

We assume now that d is equal to 1. To set the stage, we repeat some material from [8]:

> We find that the only possibility for the check matrix is the t-fold repetition of $g(S_i)$ (generator matrix of a simplex code of length $2^i - 1$) with l zero-columns appended, yielding $n = t(2^i - 1) + l$. It amounts to appending all possible tails of length l to codewords described in Proposition 5.2. It is easy to check that there are 2 kinds of covering equalities (namely, vectors coinciding with, or being at distance 1 from, codewords on the first $t(2^i - 1)$ coordinates):
>
> $$\begin{aligned} m_0 + lm_1 &+ \binom{t}{2}(2^i - 1)m_2 + \binom{l}{2} m_2 = 1 \\ tm_1 &+ (2^{i-1} - 1)\, t^2 m_2 + tlm_2 = 1. \end{aligned}$$
>
> This implies
>
> (6.1) $$t^2 - t(2^i + 1 + 2l) + (l^2 + l + 2) = 0$$
>
> which has discriminant
>
> (6.2) $$D = (2^i + 1)^2 + 2^{i+2} l - 8.$$

We get a *PMC* iff $D = x^2$ has integer solutions. For example, the values $i = 3$, $l = 3$, $t = 14$ yield the *PMC* $[101, 98]$ with $j = 644$. Of course, for $i = t$ we get $8l + 1 = x^2$ having all odd x as solutions.

Now we can characterize the solutions of $D = x^2$. We need the following result:

Proposition 6.2 $(2^{i+1} - 7)$ *is a square mod* 2^{i+2}.

Proof. Proof by induction on i. If x is a solution for some i, i.e., for $\alpha \in \mathbf{N}$, $x^2 = \alpha\, 2^{i+2} + 2^{i+1} - 7$, then for any $\beta \in \mathbf{N}$ to be chosen later on, and $i \geq 3$:

$$\begin{aligned} (x + 2^{i+1}\beta + 2^i)^2 &= x^2 + 2^{i+2}(x\beta + \alpha) + 2^{i+1}x + 2^{i+1} - 7 + 2^{2i}(1 + 4\beta^2 + 4\beta) \\ &\equiv 2^{i+2}\left(x\beta + \alpha + \tfrac{x-1}{2}\right) + 2^{i+2} - 7 \bmod 2^{i+3}. \end{aligned}$$

Since x is odd, we can certainly find β to make $x\beta + \alpha + \frac{x-1}{2}$ even. Then $x + 2^{i+1}\beta + 2^i$ is a solution for $i + 1$. For $i \leq 2$, the proposition is easily checked. □

The first proof of this proposition was given by I. Shparlinski during the present Workshop.

Obviously, the congruence

$$x^2 \equiv 2^i - 7 \pmod{2^{i+2}}$$

has 4 roots. Denoting by a the one which lies in $[0, 2^{i+1}]$, they are

$$a,\ 2^{i+1} - a,\ 2^{i+1} + a,\ 2^{i+2} - a.$$

Now direct calculations lead to the solution of (6.2), giving the possible l. Then t is derived from (6.1).

Theorem 6.1 *Linear PMC with $m_0 = m_1 = m_2 = 1/j$, $d = 1$ exist only for the following sets of parameters:*

$l = (\gamma^2 2^{2i+2} \pm 2^{i+2}\gamma a + a^2 - 2^{i+1} + 7 - 2^{2i})/2^{i+2}$

$t = (2^i + 1 + 2l \pm \sqrt{(2^i+1)^2 + 2^{i+2}l - 8})/2$

$n = t(2^i - 1) + l$

$k = n - i$

$j = (2^{i-1} - 1)t^2 + t(1 + l)$,

for $\gamma \in \mathbf{Z}$, provided $l \in \mathbf{N}$. □

6.2 The case $s = 2$

We have found the following *PMC* codes C in this case ($d = s = \delta = 2$); see [8] for constructions.

C		$C^\perp$
$[5,1;5]$	$j = 1$	$[5,4;2,4]$
$[5,2,2]$	$j = 2$	$[5,3;2,4]$
$[5,3,2]$	$j = 4$	$[5,2;2,4]$
$[10,7,2]$	$j = 7$	$[10,3;4,7]$
$[37,32,2]$	$j = 22$	$[37,5;16,22]$
$[8282,8269,2]$	$j = 4187$	$[8282,13;4096,4187]$

The first is a classical perfect code. The notation $[n,k;w_1,w_2,\ldots]$ stands for an $[n,k]$ code in which all nonzero weights are among $w_1, w_2, \ldots$. In the above codes $C^\perp$, since $s = 2$, both weights are present. All the above codes C are *PMC* codes.

Conjecture 6.1 *We conjecture the nonexistence of PMC with $d = s = \delta = 2$ other than those in the table.*

References

[1] E. F. Assmus, Jr., H. F. Mattson, Jr., R. Turyn, "Cyclic codes," Final Report, Contract no. AF 19(604)–8516, AFCRL, April 28, 1966. Sylvania Applied Research Laboratory, Waltham, Mass. (Document no. AFCRL–66–348.)

[2] L. A. Bassalygo, G. V. Zaitsev, V. A. Zinoviev, "On Uniformly Packed Codes," *Problems of Inform. Transmission*, vol. 10, no. 1, 1974, pp. 9–14.

[3] L. A. Bassalygo, G. V. Zaitsev, V. A. Zinoviev, "Note on Uniformly Packed Codes," *Problems of Inform. Transmission*, vol. 13, no. 3, 1977, pp. 22–25.

[4] A. R. Calderbank, J. M. Goethals, "On a Pair of Dual Subschemes of the Hamming Scheme $M_n(q)$, " *Europ. J. Combinatorics*, 6, 1985, pp. 133–147.

[5] R. F. Clayton, "Multiple Packings and Coverings in Algebraic Coding Theory," Thesis, Univ. of California, Los Angeles, 1987.

[6] G. D. Cohen, P. Frankl, "On tilings of the binary vector space," *Discrete Math.*, 31, 1980, pp.271-277.

[7] G. D. Cohen, M. G. Karpovsky, H. F. Mattson, Jr., J. R. Schatz, "Covering Radius—Survey and Recent Results," *IEEE Trans. Inform. Theory* IT–31, 1985, pp. 328–343.

[8] G. D. Cohen, S. N. Litsyn, H. F. Mattson, Jr., "Binary Perfect Weighted Coverings," Sequences '91, Positano, 17–22 June 1991, *Springer-Verlag Lecture Notes in Computer Science.* To appear.

[9] P. Delsarte, "Four Fundamental Parameters of a Code and Their Combinatorial Significance," *Information and Control*, vol. 23, no. 5, 1973, pp. 407–438.

[10] J. M. Goethals, S. L. Snover, "Nearly Perfect Binary Codes," *Discrete Math.*, 3, 1972, pp. 65–88.

[11] J. M. Goethals, H. C. A. van Tilborg, "Uniformly packed codes," *Philips Res. Repts* 30, 1975, pp. 9–36.

[12] M. Karpovsky, "Weight distribution of translates, covering radius and perfect codes correcting errors of the given multiplicities," *IEEE Trans. Inform. Theory* IT-27, 1981, pp. 462–472.

[13] J. E. MacDonald, "Design methods for maximum minimum-distance error-correcting codes," *IBM J. Res. Devel.* vol. 4, 1960, pp. 43–57.

[14] N. V. Semakov, V. A. Zinoviev, G. V. Zaitsev, "Uniformly Packed Codes," *Problems of Inform. Transmission*, vol. 7, 1971, no. 1, pp. 38–50.

[15] Th. Skolem, P. Chowla, and D. J. Lewis, "The Diophantine equation $2^{n-2} - 7 = x^2$ and related problems," *Proc. Amer. Math. Soc.* 10 , 1959, pp. 663–669.

[16] G. J. M. van Wee, G. D. Cohen, S. N. Litsyn, "A note on Perfect Multiple Coverings of Hamming Spaces," *IEEE Trans. Inform. Theory*, IT – 37, no. 3, 1991, pp. 678–682.

AN EXTREMAL PROBLEM RELATED TO THE COVERING RADIUS OF BINARY CODES

G. Zémor

E.N.S.T.
Dept. Réseaux
46 rue Barrault
75634 Paris Cedex 13
FRANCE

ABSTRACT

We introduce an extremal problem akin to the search for the parameters of binary linear codes of minimal distance 3 and the largest possible covering radius ; we solve the problem in some cases, give some bounds, and also show how it relates to the construction of good codes with distance 3.

1 Introduction

Denote by $\boldsymbol{F}$ the field on two elements $\{0,1\}$, and by $\boldsymbol{F}^r$ the set of binary r-tuples. $d(x,y)$ stands for the Hamming distance between two r-tuples x and y. Given a generating set S of $\boldsymbol{F}^r$, denote by $t(S)$ the smallest integer t such that :

$$\forall x \in \boldsymbol{F}^r, \exists s_1, s_2, \ldots, s_k \in S \text{ such that } x = s_1 + s_2 + \ldots + s_k, \text{ with } k \leq t.$$

Given a positive integer t, we denote by $s_r(t)$ the smallest integer s such that

$$\text{for any generating set } S \text{ of } \boldsymbol{F}^r, \quad |S| \geq s \Rightarrow t(S) \leq t.$$

In other words, $s_r(t) - 1$ is the largest cardinality of a generating set S of $\boldsymbol{F}^r$ such that $t(S) > t$. We adress the problem of determining the numbers $s_r(t)$; besides from being interesting on its own, motivation for this problem stems from a problem of writing on memories (WOMs - Write Once Memories -)

with the particular constraint that the transition "0" to "1" is allowed, but the transition "1" to "0" is impossible. The problem is to reuse the WOM as many times as possible, in other words to find ways to write on the memory which use up few $0 \rightarrow 1$ transitions. In an ingenious scheme originating in [RiS] and developped in [CGM], the number $s_r(t)-1$ represents the minimum number s needed to claim that whenever there are s or more memory positions available (containing a "0") any message of r bits can be written on the memory using at most t $0 \rightarrow 1$ transitions. For more details see [CGM], [God], [Zem].

Let us start by a very simple result on $s_r(t)$.

Proposition 1 *For $r \geq 2$ we have $s_r(2) = 2^{r-1} + 1$*

Proof : Let e be any non zero element of $\boldsymbol{F}^r$; Let H be a hyperplane of $\boldsymbol{F}^r$ not containing e. Then $S = \{0\} \cup (H+e) \setminus \{e\}$ is a generating set of $\boldsymbol{F}^r$ such that $|S| = 2^{r-1}$ and $S \cup (S+S) \neq \boldsymbol{F}^r$. Furthermore if S is such that $2|S| > 2^r$, then $S + S = \boldsymbol{F}^r$; (Recall the following version of the pigeon-hole principle, in any group G, if U and V are two subsets such that $|U| + |V| > |G|$ then $U + V = G$).

□

When $t \geq 3$, the problem is more difficult. We will use some coding theory, namely the following : to any generating set S of $\boldsymbol{F}^r$ associate the linear code $C(S)$ a parity check matrix of which has its columns equal to the nonzero elements of S ; ($C(S)$ is defined up to a permutation of coordinates). $C(S)$ is a code in length $n = |S \setminus \{0\}|$, has a minimum distance at least 3, a dimension $n-r$, and : $t(S) = \rho(C(S))$ where $\rho(C)$ denotes the covering radius of a code C, i.e.

$$\rho(C) = \min \{t \mid \forall x \in \boldsymbol{F}^r, \exists c \in C, d(x,c) \leq t\}$$

The problem of determining $s_r(t)$ can be considered as a search for the parameters of linear codes in distance 3 with a large covering radius.

2 A lower bound on $s_r(t)$

The following construction suggests itself quite naturally :

Definition : call a τ-cylinder in F^r the subset $S = B_{\tau,1}(0) \times F^{r-\tau}$, where $B_{\tau,1}(0)$ denotes the ball centered on 0 and of radius 1 in F^τ ; in other words S is the subset of vectors of F^r whose first τ coordinates make up a word of weight at most one. Obviously If S is a τ-cylinder in F^r, then $t(S) = \tau$. Furthermore we have $|S| = (\tau + 1)2^{r-\tau}$; whence, bearing in mind that $s_r(t)$ should be larger than the cardinality of a $(t+1)$-cylinder :

Proposition 2 *When* $r \geq t+1$, *we have the lower bound :* $(t+2)2^{r-t-1} < s_r(t)$

It seems to us that this lower bound is the best possible ; we conjecture

CONJECTURE : For $r \geq t+1$, $\quad s_r(t) = (t+2)2^{r-t-1} + 1$

Note that the conjecture is true for $t = 2$; (proposition 1). We will also prove it in a few other situations.

3 An upper bound on $s_r(t)$

Denote as usual by $k(n,d)$ the maximum dimension of a linear code in length n and of minimum distance at least d, and by $A(n,d)$ the maximum cardinality of a (not necessarily linear) code in length n and distance d. The following proposition appears originally in [God] and is also made use of in [ZeC], we present it here in a slightly different form.

Proposition 3 *Let* C *be a [n,k,d]-linear code, and let* ρ *be its covering radius; we have*

(i) $k + k(\rho, d) \leq k(n,d)$

(ii) $k + \log_2 A(\rho, d) \leq \log_2 A(n,d)$

Proof : Let C be a [n,k,d] code with covering radius ρ, and let z be a vector of weight ρ and such that its distance to any word of C is at least ρ. Assume without loss of generality that the support of z is $supp(z) = \{1, 2, \ldots, \rho\}$. Consider a code C' in length ρ, minimal distance d, and optimal, i.e. either with dimension $k(\rho, d)$ if it is linear (case **(i)**), or else with $A(\rho, d)$ codewords (case **(ii)**). Let $(C'|0)$ be the code in length n obtained from C' by appending

$0 \in F^{n-\rho}$ to all words in C'. It is not difficult to check that the sum $C+(C'|0)$ is a code with minimal distance at least d and with $|C||C'|$ codewords. This proves both **(i)** and **(ii)**.

□

Recall the well known

$$k(n,3) = n - 1 - \lfloor \log n \rfloor$$

Therefore if S is a generating set of F^r not containing 0 and such that $|S| = n$ and $C(S)$ has covering radius ρ, then proposition 3 **(i)** (in the case $d = 3$) yields

$$n - r + (\rho - 1 - \lfloor \log \rho \rfloor) \leq n - 1 - \lfloor \log n \rfloor$$

in other words

$$t(S) - \lfloor \log t(S) \rfloor \leq r - \lfloor \log |S| \rfloor$$

Hence, we have the following upperbound on $s_r(t)$

Proposition 4 $s_r(t) \leq 2^{r-t+\lfloor \log(t+1) \rfloor} + 1$

Notice that when $t = 2^m - 2$ for any integer $m \geq 2$, the lower bound and upperbound of propositions 2 and 4 meet, i.e.

Proposition 5 *The conjecture is true for* $t = 2^m - 2$ *for any integer* $m \geq 2$.

4 Other results

4.1 $r \leq t+2$

When r is not too much bigger than t, determining $s_r(t)$ becomes easier, for instance it is clear that in F^{t+1}, the only generating subsets of F^r that satisfy $t(S) \geq t+1$ are its bases, perhaps enriched by the zero vector : they contain at most $t+2$ elements, hence :

$$s_{t+1}(t) = t + 3$$

Next we determine $s_{t+2}(t)$ by an "ad hoc" method.

Proposition 6 $s_{t+2}(t) = 2(t+2) + 1$

Proof : Let S be a subset of F^{t+2}, with $t(S) = t+1$, and of maximum cardinality. Let x be an element of F^{t+2} that is not a sum of t or less elements of S. x can be expressed as the sum

$$x = s_1 + s_2 + \ldots + s_{t+1} \tag{1}$$

where the s_i are all elements of S. Note that $(s_i)_{1\leq i\leq t+1}$ is necessarily an independant family of vectors of S. We can therefore choose a basis B of F^{t+2} made up of $(s_i)_{1\leq i\leq t+1}$, and some additional element that we will denote by σ. Write that S can be partitioned as follows :

$$S = \{0\} \cup B \cup E$$

We need to prove that E contains at most $t+1$ elements.

Fact 1 : Let e be an element of E. Relative to the base B, the σ-coordinate of e is 1.

If not, then e is generated by $(s_i)_{1\leq i\leq t+1}$, i.e. e can be expressed as $e = \Sigma_{j\in J}\, s_j$, with $|J| \geq 2$ since $e \notin B$. But then $x = e + \Sigma_{j\in[1,t+1]\setminus J}\, s_j$ contradicts the minimality of the expression (1).

Fact 2 : The weight of e (relative to B) is at most 3

Otherwise x can be expressed as $x = e + \sigma + \Sigma_{j\in[1,t+1]\setminus J}\, s_j$ where $|J| \geq 3$ which contradicts the minimality of (1).

Fact 3 : If $e' \in E$, and $e' \neq e$ then $d(e, e') \leq 2$ (where $d(,)$ denotes the Hamming distance relative to B). Hence if e and e′ are elements of weight 2 and 3 relative to B, then supp(e) ⊂ supp(e′

Otherwise $x = e + e' + \Sigma_{j\in[1,t+1]\setminus J'}\, s_j$ again contradicts the minimality of (1).

Using facts 1 to 3, it is not difficult to establish that

Fact 4 : If $|E| > 3$, and E contains an element of weight 3, then all elements of E have a 1 in a common coordinate different from the σ-coordinate. In other words ($w(\,)$ denoting the weight relative to B)

$$\exists u \in [1, t+1] \text{ such that } \forall e \in E,\ \ w(s_u + e) < w(e)$$

The case $|E| = 3$ can be treated separately, since it corresponds to the easy case $t = 2$ which is already solved. We conclude in the following way, supposing $|E| > 3$:

1. Either all elements of E are of weight two, and then all elements of $\sigma+E$ are of weight one (fact 1), hence $|E| \leq t+1$.

2. Or E contains elements of weight 3, but by fact 5 all the elements of $\sigma + s_u + E$ are of weight one and therefore $|E| \leq t+1$.

□

Remark : in the above proof, S not only has the cardinality of a $(t+1)$-cylinder, but it is necessarily a $(t+1)$-cylinder (when $t \geq 3$) : (relative to the base $B = (s_1, s_2, \ldots, s_{t+1}, \sigma)$ in case 1, and relative to the base $(s_1, s_2, \ldots, s_{t+1}, \sigma + s_u)$ in case 2).

It is not clear to us whether this line of reasoning can be generalised to larger r.

4.2 The case $t = 3$

As we have seen, the bound of proposition 4 is sharpest when $t = 2^m - 2$. It is least informative when $t = 2^m - 1$. When $m = 2$ i.e. $t = 3$, proposition 4 becomes $s_r(3) \leq 2^{r-1}+1$, which is really a trivial bound since $2^{r-1}+1 = s_r(2)$. The case $t = 3$ is therefore the case when coding-theoretic arguments seem the least effective, and another line of reasoning should be brought in.

The case of small r can be checked by (slightly improved) exhaustive search. It is proved in [Zem] that the conjecture holds for $t = 3$ and $r \leq 7$; we will not reproduce the proofs here, for though simple, they are tedious and uninformative. For larger r we adopt an approach that can be described informally as follows : if a subset S of $\boldsymbol{F}^r$ satisfies $S + S + S \neq \boldsymbol{F}^r$ then $|S+S| + |S| \leq |\boldsymbol{F}^r|$ (by the pigeon-hole principle). We will therefore look for a characterisation of those subsets S such that $|S+S|$ is "small".

A general theorem in abelian groups about pairs of sets S and T such that $|S+T|$ is "small" is the following theorem of Mann [Man] :

Theorem 4.1 *Let S be a subset of an abelian group G. One of the following holds :*

- *There exists a non trivial subgroup H of G such that*

$$|S+H| - |S| < |H| - 1$$

- *For any subset T of G such that $S+T \neq G$ we have*

$$|S+T| \geq |S| + |T| - 1$$

Examples of groups for which this theorem cannot be improved are easily found ; this is not the case of $\boldsymbol{F}^r$ however, in [Zem] we managed to prove the following extension :

Theorem 4.2 *Let S be a subset of $\boldsymbol{F}^r$. Let k be a nonnegative integer. One of the following holds:*

- *There is a nontrivial subgroup H of $\boldsymbol{F}^r$ such that*

$$|S+H| - |S| < |H| + k$$

- *For any subset T of $\boldsymbol{F}^r$ such that $k \leq |T|^2 - 2$ and $2 \leq |\boldsymbol{F}^r| - |S+T|$ we have*

$$|S+T| \geq |S| + |T| + k$$

This theorem can be applied to yield :

Proposition 7 $s_r(3) \leq 2^r/3 + 1$.

Proof : We will prove that if S is a generating set of $\boldsymbol{F}^r$ such that $|S| > |\boldsymbol{F}^r|/3$ then

$$S+S+S = \boldsymbol{F}^r$$

First check the result for $r \leq 4$. We have $s_1(3) = 1, s_2(3) = 2, s_3(3) = 3, s_4(3) = 6$. Then proceed by induction ; suppose this result holds for all r', $r' < r$. Suppose S is a generating subset of $\boldsymbol{F}^r$ such that $|S| > |\boldsymbol{F}^r|/3$.

1. Either $|S+S| \geq 2|S|$, and then $|S| + |S+S| > |\boldsymbol{F}^r|$ and the pigeon-hole principle implies $S+S+S = \boldsymbol{F}^r$

2. Or $|S+S| < 2|S|$, in which case theorem 4.2 tells us that there is a nontrivial subgroup H of $\boldsymbol{F}^r$ such that

$$|S+H| - |S| \leq |H| - 1. \tag{2}$$

   Suppose S_i and S_j are each a nonempty intersection of S with a coset modulo H. Then it is easily checked that (2) and the pigeon-hole principle imply that $S_i + S_j$ is exactly a coset modulo H. Besides, since S contains more than one coset modulo H (it is a generating set), there is one S_i such that $|S_i + H| - |S_i| < |H|/2$ i.e. such that $|S_i| > |H|/2$ and hence $S + S \supset S_i + S_i \supset H$. All this means that

$$S + S + H = S + S$$

in other words $S + S$ is a union of cosets modulo H. Denote by S/H the set of those cosets ; since necessarily $|S/H| > |\mathbf{F}^r/H|/3$, by our induction hypothesis

$$S/H + S/H + S/H = \mathbf{F}^r/H.$$

S being a union of cosets modulo H, this implies

$$S + S + S = \mathbf{F}^r$$

□

Note that the upper bound on $s_r(3)$ of proposition 7 is quite close to the lower bound of proposition 2, even though there is still a small gap.

4.3 Relations between $s_r(t)$ and $A(n,3)$

We have used proposition 3 to obtain an upper bound on $s_r(t)$, but clearly the same construction can be used to obtain large codes in distance 3 provided the two codes in the direct sum in the proof of proposition 3 are large enough. In itself this remark is insufficient to actually improve known lower bounds on $A(n,3)$ but by studying a table of known bounds on $A(n,3)$ (e.g. in [McS]) we can state the following, which, in a way provides further motivation for the study of $s_r(t)$

Proposition 8 *One of the following holds*

- *Either the conjecture holds for* $t = 7, t = 8,$ *or* $t = 10$
- *Or there exist for values of* n *of the form* $s_r(7) - 1$, *or* $s_r(8) - 1$, *or* $s_r(10) - 1$ *codes of length* n *and distance 3 whose cardinality exceeds that of the best known codes.*

5 A summary of results and concluding remarks

We were able to prove in the general case that when $t+1 \leq r$,

$$(t+2)2^{r-t-1} < s_r(t) \leq 2^{r-t+\lfloor \log(t+1) \rfloor} + 1$$

For small t we have

$$\text{for } r \geq 2, \qquad s_r(2) = 2^{r-1} + 1,$$

$$5 \times 2^{r-4} < s_r(3) \leq 2^r/3 + 1$$

and for $t+1 \leq r \leq t+2$ we have

$$s_r(t) = (t+2)2^{r-t-1} + 1$$

We wish to point out the following : for $t \geq 4$ we have not met any generating set S of $\boldsymbol{F}^r$ that has both the same cardinality and the same $t(S)$ (i.e. t) as a t-cylinder, and that is not a t-cylinder relative to any basis of $\boldsymbol{F}^r$. It should perhaps be conjectured that there are none ; that is not the case however for $t = 3$ since the set S in the proof of proposition 1 verifies $|\boldsymbol{F}^r \setminus (S+S)| = 1$ while a 3-cylinder T verifies $|\boldsymbol{F}^r \setminus (T+T)| = 2^{r-3}$.

References

[CGM.83] G.D. Cohen, P. Godlewski, and F. Merkx, "Linear Block Codes for Write-Once Memories", IEEE Trans. on Inf. Theory, IT 29, 3, (1983) 731-739.

[God.87] P. Godlewski, "WOM-codes construits à partir des codes de Hamming", *Discrete Math.*, 65 (1987), 237-243.

[McS.77] F.J. Mac Williams and N.J.A. Sloane, *"The Theory of Error-Correcting Codes"* 1977 North Holland Amsterdam.

[Man.65] H.B. Mann, *"Addition theorems"*, 1965 Wiley New-York.

[RiS.82] R.L. Rivest and A. Shamir, "How to reuse a "Write Once" Memory", Inf. and Control 55, (1982), 1-19.

[Zem.89] G. Zémor, "Problèmes combinatoires liés à l'écriture sur des mémoires" Phd dissertation, Nov. 1989.

[ZeC.91] G. Zémor and G. Cohen, "Error-Correcting WOM-codes", IEEE Trans. on Inf. Theory, IT 37, 3, (1991) 730-735.

BOUNDS ON COVERING RADIUS OF DUAL PRODUCT CODES

G.L.Katsman

Institute of Control Science
Profsoyuznaia 65, Moscow, USSR

We present some new lower and upper bounds for the covering radius of codes dual to the product of parity check codes.

1. Introduction

Let V_1 be an $[n_1,k_1]$ code, V_2 be an $[n_2,n_2-1]$ parity check code, W $[n_1n_2,n_2(n_1-k_1)+k_1]$ be the dual of the iterative code $V_1 \otimes V_2$. Our problem will be to estimate the covering radius ρ_W of W. A similar construction called the extended direct sum was studied by Graham and Sloane [1]. They obtained an upper bound for ρ_W.

For a parity check $[n_1,n_1-1]$ code V_1 $(n_1 \leqslant n_2)$ denote by $\rho_W \triangleq \rho(n_1, n_2)$. In this case our problem is a simple generalization of the Berlekamp - Gale game [2].

In [3] a proposition that gives a method for estimating ρ_W from below, was presented. Here we discuss some corollaries of this proposition. In Section 3 , we give an upper bound for $\rho(n_1,n_2)$ which is sometimes better than the bound of [1].

2. Lower bounds

In [3], the following proposition was proved.

Proposition 1. Let $V_1^{\perp}$ be the dual of V_1 , Ω be an $[n_1,\gamma]$ code, $V_1^{\perp} \subset \Omega$, $n_2 = 2^{\gamma-(n_1-k_1)}$, $\Omega = \Omega_0, \ldots, \Omega_{2^{n_1-\gamma}-1}$ be cosets of Ω, ω_{ij} $(i = 0, 1, \ldots, 2^{n_1-\gamma}-1,\ j = 1, \ldots, n_2)$ be coset leaders of the code $V_1^{\perp}$ in the code Ω_i.

Then

$$\rho_W \geq \min_i \sum_{j=1}^{n_2} wt(\omega_{ij})$$

Sketch of the proof. The vectors of the code W are $n_2 \times n_1$ matrices. Let M be a $n_2 \times n_1$ matrix which rows are $\omega_{01}, \ldots, \omega_{0n_2}$. It is not difficult to see that the weight of coset leader M + W is $\min_i \sum_{j=1}^{n_2} wt(\omega_{ij})$

Corollary 1.

$$\rho(n, 2^{n-2}) \geq \begin{cases} \frac{n}{2}[2^{n-2} - \binom{n-2}{(n-2)/2}], & n \text{ even} \\ (n-1)[2^{n-3} - \binom{n-3}{(n-3)/2}] + 2^{n-3}, & n \text{ odd} \end{cases}$$

Proof. Let n be even, $n_1 = n$, let Ω be a parity check [n,n-1] code. In this case V_1 is an [n,n-1] code, $V_1^\perp$ is an [n,1] code, and $V_1^\perp \subset \Omega$, $\gamma = n-1$, and $n_2 = 2^{n-2} = 2^{\gamma-(n_1-k_1)}$. It is clear that Ω_0 is the set of all even weight vectors of length n , and Ω_1 is the set of all odd weight vectors of length n.

Suppose $\Omega_i' = \{\omega_{i1}, \ldots, \omega_{i2^{n-2}}\}$, $i = 0,1$, A_n is a set of vectors of length n and weight n/2, $|A_n| = \binom{n}{n/2}/2$, $\forall\ a_i, a_j \in A_n$, $a_i \neq \bar{a}_j$, $B_{n,0}$ - is the set of vectors of length n and of even weight not greater than $\frac{n}{2} - 1$, and $B_{n,1}$ is the set of vectors of odd weight not greater than $\frac{n}{2} - 1$. Now, if $n = 4k$ then $\Omega_0' = B_{n,0} \cup A_n$ and $\Omega_1' = B_{n,1}$, and if $n = 4k+2$ then $\Omega_0' = B_{n,0}$ and $\Omega_1' = B_{n,1} \cup A_n$.

Now from proposition 1 it is not difficult to see that

$$\rho(n, 2^{n-2}) \geq \frac{n}{2}[2^{n-2} - \binom{n-2}{(n-2)/2}]$$

For n odd, Ω is the [n,n-1] code with single check (11...10). The proof is similar to the previous case.

Observe that from the upper bound for $\rho(n_1,n_2)$ presented in [1], one easily obtains

$$\rho(n,\ 2^{n-2}) \leqslant \frac{n}{2}[2^{n-2} - \binom{n-1}{(n-1)/2} / 2\ ] \quad .$$

So, we have $\rho(5,8)=12$, $30 \leqslant \rho(6,16) \leqslant 33$, $76 \leqslant \rho(7,32) \leqslant 77$.

Corollary 2 For $n=2^m-1$, we have

$$\rho(n,\ 2^{n-m-1}) \geqslant \frac{n[2^{n-1} - \binom{n-1}{(n-1)/2}] - 2\binom{(n-3)/2}{(n-3)/4}}{2(n+1)}$$

Sketch of the proof. In this case let Ω be the $[2^m-1,\ 2^m-m-1]$ Hamming code, $n_1 = n = 2^m-1$, $n_2 = 2^{n-m-1}$, $V_1^{\perp}$ be the [n,1] code, and $V_1^{\perp} \subset \Omega$. Define $S_i = \sum_{j=0}^{n_2} wt(\omega_{ij})$. At first, we calculate S_0. This is not difficult because Ω_0 is the Hamming code with known weight distribution, and $\{\omega_{01}, \ldots, \omega_{0n_2}\}$ is the set of all vectors of the Hamming code with weight not greater than $n_1/2$. We have

$$S_0 = \frac{\sum_{i=0}^{(n-1)/2} i\binom{n}{i} + n\binom{(n-3)/2}{(n-3)/4}}{n+1}$$

All cosets Ω_1 , ... , Ω_{2^m-1} have equal weight distribution. So $S_1 = S_2 = \ldots = S_{2^m-1} = S$. On the other hand, it is easy to see that

$$\sum_{i=0}^{2^m-1} S_i = \sum_{i=0}^{(n-1)/2} i\binom{n}{i} .$$

So

$$S = \frac{\sum_{i=0}^{(n-1)/2} i\binom{n}{i} - \binom{(n-3)/2}{(n-3)/4}}{n+1} \text{, and } S < S_0 .$$

Finally,

$$\sum_{i=0}^{(n-1)/2} i\binom{n}{i} = \frac{n[2^{n-1} - \binom{n-1}{(n-1)/2}]}{2} \quad \text{and}$$

$$\rho(n,2^{n-m-1}) \geqslant S = \frac{n[2^{n-1} - \binom{n-1}{(n-1)/2}] - 2\binom{(n-3)/2}{(n-3)/4}}{2(n+1)}$$

Observe that the upper bound presented in [1], implies

$$\rho(n,2^{n-m-1}) \leqslant \frac{n[2^{n-1} - \binom{n-1}{(n-1)/2}]}{2(n+1)}$$

So we have $\rho(7,8)=19, 6070 \leqslant \rho(15,1024) \leqslant 6071$.

3. Upper bounds

In this section, we present a new upper bound for $\rho(n_1,n_2)$. This bound is sometimes better then the bound from [1]. Moreover, the bound from [1] (for the case of a parity check code V_1) can be obtained from this new bound.

Proposition 2. Let $l_0^*, l_1^*, \ldots, l^*_{\lceil n_1/2 \rceil}$ be the solution of the integer linear programming problem

$$\sum_{j=1}^{\lceil n_1/2 \rceil} j l_j \to \max$$

$$\sum_{j=0}^{\lceil n_1/2\rceil} l_j = n_2$$

$$\sum_{j=0}^{\lceil n_1/2\rceil} B_{ij} l_j \geqslant 0 \qquad i = 1,\dots, \lceil n_1/2\rceil$$

where

$$B_{ij} = \sum_{s=0}^{n_1} (\xi - j) \binom{j}{(j+i-s)/2} \binom{n-j}{(i+s-j)/2} ,$$

$$\xi = \min(s, n_1 - s).$$

Then $\rho(n_1,n_2) \leqslant \sum_{j=1}^{\lceil n_1/2\rceil} j l_j^* \; .$

Sketch of the proof. The vectors of the code W are $n_2 \times n_1$ matrices. Let us call matrices that have either all - one or all - -zero columns (resp., rows), matrices of type 1 (resp., of type 2). All matrices of types 1 and 2 belong to the code W. So, if for a matrix M, wt(M) = $\rho(n_1, n_2)$, then any adding of matrices of type 1 or 2 does not decrease the weight of M.

Let us add some matrix N of type 1 to M. Then in order to make all row weights at most $\lceil n_1/2\rceil$ add a matrix of type 2. Denote the resulting matrix by M'.

If the matrix M has l_j rows of weight j (j=0, 1,...,$\lceil n_1/2\rceil$), $wt(M) = \sum_{j=1}^{\lceil n_1/2\rceil} j l_j$, $\sum_{j=0}^{\lceil n_1/2\rceil} l_j = n_2$, then we can compute the average weight of the matrices M' over all posible choices of the matrix N of fixed weight (fixed number of nonzero columns). This average value must be at least wt(M).

It is easy to see that the upper bound presented in [1] can be obtained if we compute the average value of wt(M') over all possible choices of the matrices N. So the upper bound from [1] follows from the new bound.

Example 1. For $n_1 = 5$, from Proposition 2 we obtain an upper bound for $\rho(n_1,n_2)$.

If l_1^*, l_2^* is the solution of the following integer linear programming problem

$$l_1 \geqslant 0,\ l_2 \geqslant 0,$$
$$l_1+l_2 \leqslant n_2$$
$$2l_1+7l_2 \leqslant 5n_2$$
$$14l_1+25l_2 \leqslant 20n_2;$$

Then $\rho(5,n_2) \leqslant l_1^*+2l_2^*$

For $n_2 = 9$, from this bound we obtain $\rho(5,9) \leqslant 13$. From the upper bound from [1] we have $\rho(5,9) \leqslant 14$. It is not difficult to see that $\rho(5,9) = 13$.

Example 2. From the new upper bound it follows that $\rho(7,32) \leqslant 76$, and from the lower bound of Section 2 we have $\rho(7,32) = 76$.

References

1. Graham R.L., Sloane N.J.A. On the Covering Radius of Codes. IEEE Trans. Inform. Theory, 1985, IT-31, p. 385-401.
2. Fishburn P.C., Sloane N.J.A. The Solution of Berlekamp's Switching Game. Discrete Math. 1989, 74, p. 263-290.
3. Katsman G.L. Covering Radius of Codes Being Dual to Iterative Ones. Proceedings of Fifth Joint Soviet - Swedish International Workshop on Information Theory. Convolutional Codes; Multi-User Communication. Jan. 13-19, 1991 Moscow, USSR, p. 91-92.

Remarks on Greedy Codes

VERA PLESS
Mathematics Department
University of Illinois at Chicago
Chicago, IL 60680

Greedy codes [1] are families of binary codes constructed by a greedy algorithm. As the lexicodes of Conway and Sloane [2] are greedy codes we start by describing them.

The first vector placed in a lexicode of length n and minimum weight d is the zero vector. The other vectors are determined by the following greedy algorithm. Consider the 2^n binary n-tuples as numbers in their binary representation and list them in lexicographic order. The second vector placed in the lexicode is the first vector of weight d which occurs in our list. After i vectors in the code are chosen the $i+1^{st}$ vector is the next vector on the list which has distance d or more from the vectors already chosen. Conway and Sloane prove that this set of vectors form a linear code of length n and minimum weight d. This code is a lexicode and the theory of lexicodes is related to the Sprague-Grundy theory of impartial games. The lexicodes are shown to be linear by relating their vectors to winning moves in a certain impartial game [2]. No information is given about the dimension of the code constructed except in special cases. When the dimension was computed, by constructing the code, it turned out to be very good.

To illustrate the procedure for constructing lexicodes we list the $2^5 = 32$ binary vectors of length 5 in lexicographic order and construct the lexicode for $n = 5$ and $d = 3$.

Research partially supported by NSA Grant MDA 904-91-H-0003

The 2^5 vectors	g-values
00000	0 √
00001	1
00010	2
00011	3
00100	3
00101	2
00110	1
00111	0 √
01000	4
01001	5
01010	6
01011	7
01100	7
01101	6
01110	5
01111	4
10000	5
10001	4
10010	7
10011	6
10100	6
10101	7
10110	4
10111	5
11000	1
11001	0 √
11010	3
11011	2
11100	2
11101	3
11110	0 √
11111	1

The four checked vectors constitute the lexicode for $n = 5$ and $d = 3$ which we see has dimension 2. These are also the vectors with g-value zero, but we have not yet explained what the g-values are.

We [1] wondered what would happen if we wrote the numbers in their Gray-code order, instead of their lexicographic order, and used the same algorithm. We noticed that the set of vectors chosen in this way is also linear! We demonstrate this for $n = 5$ and $d = 3$ again in Table 1 [1]. We still have g-values in the table and other new things, the y_i. The code obtained in this way has dimension 2 and is called the Gray-greedy code of length 5 for $d = 3$.

Gray order of F_2^5	g-values	g-values as vectors	Gray-greedy code
00000	0	000	♠
$y_1 =$ 00001	1	001	
$y_2 =$ 00011	2	010	
00010	3	011	
$y_3 =$ 00010	1	001	
00111	0	000	♠
00101	3	011	
00100	2	010	
$y_4 =$ 01100	4	100	
01101	5	101	
01111	6	110	
01110	7	111	
01010	5	101	
01011	4	100	
01001	7	111	
01000	6	110	
$y_5 =$ 11000	1	001	
11001	0	000	♠
11011	3	011	
11010	2	010	
11110	0	000	♠
11111	1	001	
11101	2	010	
11100	3	011	
10100	5	101	
10101	4	100	
10111	7	111	
10110	6	110	
10010	4	100	
10011	5	101	
10001	6	110	
10000	7	111	

Table 1: Gray-greedy code of designed distance 3.

We let $\oplus$ denote addition of binary vectors. If a and b are regarded as numbers written in binary, $a \oplus b$ is called the nim-sum of a and b. For example

$$12 \oplus 5 = 9$$
$$(1,1,0,0) \oplus (0,1,0,1) = (1,0,0,1)$$

Let $\mathcal{B}$ denote an ordered basis $y_1, \ldots, y_n$ of F_2^n. This is just a basis of F_2^n written in a specific order. Using $\mathcal{B}$ we can list the vectors in F_2^n in order as follows:

$$\begin{array}{c} 0 \\ y_1 \\ y_2 \\ y_2 \oplus y_1 \\ y_3 \\ y_3 \oplus y_1 \\ y_3 \oplus y_2 \\ y_3 \oplus y_2 \oplus y_1 \\ y_4 \\ \vdots \end{array}$$

If $\mathcal{B}$ is the standard unit basis in the order

$$e_1 = (0, \ldots, 0, 1),\ e_2 = (0, \ldots, 1, 0), \ldots, e_n = (1, 0, \ldots, 0),$$

then the vectors in F_2^n are listed in the standard lexicographic order of binary n-tuples.

If $\mathcal{B}$ is the following ordered basis

$$(0, \ldots, 0, 1), (0, \ldots, 0, 1, 1), (0, \ldots, 0, 1, 1, 0), \ldots, (1, 1, 0, \ldots, 0)$$

then the vectors in F_2^n are listed in their Gray-code order.

Let $\mathcal{B}$ be an ordered basis of F_2^n and let d be an integer with $0 \leqq d \leqq n$. Applying the Greedy algorithm (for d) to the order of F_2^n given by $\mathcal{B}$ we obtain a code C whose minimum distance is at least d. This is the $\mathcal{B}$-*greedy code* of length n and designed distance d. Even though we call the set of vectors constructed by the greedy algorithm a code C we do not as yet know that C is a linear code.

We can show that the greedy codes are linear, but in order to do so we need to define those numbers, the g-values of all the vectors. Consider the vectors listed in the order given by an ordered basis.

$$0, z_2, \ldots, z_{2^n}.$$

Then we let $g(0) = 0$. Suppose $g(z_2), \ldots, g(z_{i-1})$ have been defined. Then z_i is assigned g-value equal to the smallest non-negative integer t such that z_i has distance at least d to all vectors which are already labelled t. If no such exists, z_i is given the smallest integer not yet assigned. The greedy code C of designed distance d equals the set of all vectors whose g-value equals 0. This is equivalent to the previous definition of C by the greedy algorithm.

Using the order of the vectors in F_2^n given by an ordered basis and the g-values as defined above we are able to prove the following theorem without resorting to game theory.

Theorem 1 [1]. The greedy code C is linear.
(It clearly has length n and minimum distance $\geq d$).

Further, a parity check matrix for C is

$$H = \left[g(e_n)\ldots g(e_2)g(e_1)\right].$$

We call this the ***g*-parity check matrix** of C.

For each z in F_2^n, $g(z)$ is the syndrome of z relative to H.

We illustrate this theorem with our two examples. In the lexicode example we note that $g(e_i) = i, i = 1, 2, \ldots, 5$. Hence the g-parity check matrix for the lexicode with $n = 5$, $d = 3$, is

$$H = \begin{pmatrix} 1 & 1 & 0 & 0 & 0 \\ 0 & 0 & 1 & 1 & 0 \\ 1 & 0 & 1 & 0 & 1 \end{pmatrix}$$

It can be shown then that a parity check for the lexicode of length n with $d = 3$ is $H = (n \ldots 1)$ where these numbers are written out in binary.

By noting the g-values for the $e_i, i = 1, \ldots, 5$ in the construction of the Gray-Greedy code of length 5 and $d = 3$ it is easy to see that the following is a g-parity check matrix for this code.

$$H = \begin{pmatrix} 1 & 1 & 0 & 0 & 0 \\ 1 & 1 & 1 & 1 & 0 \\ 1 & 0 & 0 & 1 & 1 \end{pmatrix}$$

Note that the vector $(0, 1, 0, 1, 1)$ has g-value 4 which is exactly its syndrome with respect to H. Again it is interesting to notice that the columns of H are the first five numbers written in Gray-code order. Indeed, for any n the columns of the g-parity check matrix for the Gray-greedy code of length n, $d = 3$ consist of the first n numbers in Gray-code order written from right to left.

We can say more about *triangular-greedy* codes. In this situation

$$\begin{bmatrix} y_1 \\ y_2 \\ \vdots \\ y_n \end{bmatrix} = \begin{bmatrix} 0 & 0 & \ldots & 0 & 0 & 1 \\ 0 & 0 & \ldots & 0 & 1 & * \\ 0 & 0 & \ldots & 1 & * & * \\ \vdots & & & & & \\ 1 & * & \ldots & * & * & * \end{bmatrix}$$

has a triangular shape.

Lexicodes and Gray-greedy codes are triangular-greedy codes. If

$$\begin{bmatrix} y_1 \\ y_2 \\ \vdots \\ y_n \end{bmatrix} = \begin{bmatrix} 0 & 0 & \ldots & 0 & 0 & 1 \\ 0 & 0 & \ldots & 0 & 1 & 1 \\ 0 & 0 & \ldots & 1 & 1 & 1 \\ \vdots & & & & & \\ 1 & 1 & \ldots & 1 & 1 & 1 \end{bmatrix}$$

the greedy code obtained from this order is a triangular-greedy code called the ***complementary-greedy code***.

Clearly a triangular-greedy code of designed distance d has minimum weight d.

Theorem 2 [1]. If d is even, all vectors in a triangular greedy code have even weight.

The next theorem is derived from a more general theorem in [1].

Theorem 3. If d is odd,

lexicodes of length $n+1$ for $d+1$ are the extended codes of lexicodes of length n for d

Gray-greedy codes of length $n+1$ for $d+1$ are the extended codes of Gray-Greedy codes of length n for d

complementary-greedy codes of length $n+1$ for $d+1$ are the extended codes of complementary-greedy codes of length for n for d.

Theorem 4 [1]. If $d = 2$, any triangular-greedy code of length n is the $n-1$ dimensional space of all even weight vectors.

If $d = 4$, any triangular-greedy code of length n is an extended Hamming code or a shortened such code.

We also have a lower bound on the dimension of triangular-greedy codes!

Theorem 5 [1]. Let n and d be integers with d even and $4 \leqq d \leqq n$. Let C be a triangular-greedy code of length n and minimum distance d.

If $d \leqq n < 3d/2$, $\dim C = 1$.

If $n = 3d/2$, $\dim C = 2$.

If $n > 3d/2$;

$$\dim C = n - 2 - \lfloor \log_2(n-1) \rfloor \quad \text{if} \quad d = 4$$

$$\dim C \geqq \begin{cases} \lfloor \frac{4n-d-12}{2d-4} \rfloor & \text{if } d \equiv 0 \pmod 4,\ d \neq 4, 8 \\ \lfloor \frac{n}{3} \rfloor & \text{if } d \equiv 8 \quad \text{and } n > 18 \\ \lfloor \frac{4n-d-14}{2d-4} \rfloor & \text{if } d \equiv 2 \pmod 4 \end{cases}$$

(If $d = 8$ and $n \leqq 18$, exact values of $\dim C$ are given in Table 2).

We have a recursive algorithm described in [1] for constructing a g-parity check matrix for a triangular-greedy code C_n of length n. $H_1 = [1]$ is the g-parity check matrix for C_1. Suppose $H_i = [h_i \dots h_1]$ of size $m_i \times i$ has been constructed for C_i (whose columns are the g-values of the unit vectors). Our algorithm gives a constructive procedure for calculating h_{i+1} which does not require knowing he g-values of any vectors. For simplicity we only describe this algorithm for the lexicodes and the ray-greedy codes.

For the lexicodes h_{i+1} equals the smallest number $\beta \neq 0$ so that β is not a linear combination of -2 or fewer of the h_i's in H_i.

For Gray-greedy codes, $h_{i+1} = h_i + \beta$ where β is the smallest number $\neq h_i$ so that $h_i + \beta$ is not a near combination of $d-2$ or fewer of the h_i's in H_i. If $d = 3$, this is an apparently new way to generate he numbers in Gray-code order.

Using this algorithm Jesse Nemoyer computed the values given below in Table 2 [1]. Numbers in ound brackets are the dimensions of the lexicode when they differ from those for the Gray-greedy code, nd the numbers in square brackets ar those for the complementary-greedy code when they differ from hose for the Gray-greedy code.

There are several places where the Gray-greedy codes are better (that is have larger dimension) than ither the lexicodes or complementary greedy codes and one place where the complementary greedy-code better than either the Gray-greedy code, or lexicode of that length. There is also one place where he lexicode is best. The Gray-greedy code, complementary-greedy code and lexicode of length 24 and istance 8 have dimension 12 and so are each equivalent to the extended Golay code.

We note that the dimensions of the triangular-greedy codes computed in Table 2 are progressively etter than the bound in Theorem 5. Comparing Table 2 with the tables in Verhoeff [4] we see that ray-greedy codes are surprisingly good, that they either have the dimension of the best code (for n nd d) or are within one of the best code!

n : d	4	6	8	10	12
4	1	0	0	0	0
4	1	0	0	0	0
5	1	0	0	0	0
6	2	1	0	0	0
7	3	1	0	0	0
8	4	1	1	0	0
9	4	2	1	0	0
10	5	2	1	1	0
11	6	3	1	1	0
12	7	4	2	1	1
13	8	4	2	1	1
14	9	5	3	1	1
15	10	6	4	2	1
16	11	7	5	2	1
17	11	8[7]	5	2	1
18	12	9[8]	6	3	2
19	13	9	7	3	2
20	14	10	8	4	2
21	15	11	9	5	3
22	16	12	10	5	3
23	17	13(12)	11	6	4
24	18	13	12	7(6)[6]	5
25	19	14	12	7	5
26	20	15	12	8	6
27	21	16	12	9	7
28	22	17	13	9	7
29	23	18	13	10	8
30	24	19	14	11	8
31	25	20(19)[19]	15	12	9
32	26	20	16	12	10
33	26	21	16	13	10(11)[11]
34	27	22	17	14	
35	28	23	18	14	
36	29	24	19	15	
37	30	25	20	16	
38	31	26	21	17	
39	32	27	22	17[18]	
40	33	28(27)	23	18	

Table 2: Dimensions of Gray greedy codes compared to lexicodes and complementary greedy codes.

We thank G. Kabatyanski for pointing out Levenshtein's results, reference 3. This paper contains the first proof of the linearity of lexicodes. It also states, but does not prove, that $\mathcal{B}$-greedy codes are linear.

REFERENCES

1. R. A. BRUALDI AND V. PLESS, Greedy codes, preprint.
2. J. H. CONWAY AND N. J. A. SLOANE, Lexicographic codes; Error-correcting codes from game theory *IEEE Trans. Inform. Theory*, vol IT-**32**, 1986, 337–348.
3. V. I. LEVENSHTEIN, A class of systematic codes. *Dokl. Akad. Nauk SSSR* **131** (1960), 1011-1014 (Russian); translated as *Soviet Math. Dokl.* **1**, 368-371.
4. T. VERHOEFF, An updated table for minimum-distance bounds for binary linear codes, *IEEE Trans. Inform. Theory*, vol IT-**33**, 1987, 665–680.

ON NONBINARY CODES WITH FIXED DISTANCE

ILYA I. DUMER

Institute for Problems
of Information Transmission
Ermolovoy 19, Moscow GSP-4
101447 U.S.S.R.

Abstract. We consider nonbinary codes with growing length and fixed distance. Linear double-error correcting codes over an arbitrary field $GF(q)$ with length $n = q^m$ and $r \leqslant 2(m+1) + \lceil m/3 \rceil$ check symbols are constructed. Codes with greater distances are also studied.

1. INTRODUCTION

Let us consider a q-ary code $V(q,n,d)$ of length n, with Hamming distance $d > 3$ and redundancy $r(V) = n - \log_q |V|$, where $|V|$ denotes the cardinality of the code. We are interested in the minimum redundancy $\rho(q,n,d) = \min r(V)$ over all $V(q,n,d)$-codes, when q and d are fixed and n goes to infinity: $n \to \infty$. The Hamming upper bound on the cardinality $|V|$ gives for any code $V(q,n,d)$ the redundancy $r(q,n,d) \gtrsim h(q,n,d) = t \log_q n$ for $n \to \infty$, where $t = \lfloor (d-1)/2 \rfloor$.

The BCH code $B(q,n,\delta)$ of primitive length $n = q^m - 1$, $m = 2,3,\ldots$ and designed distance δ over the Galois field $L = GF(q)$ has the minimum redundancy $r(B) = \min(r_1, r_2)$ for $n \to \infty$, where $r_1 = \lceil (\delta-2)(q-1)/q \rceil m + 1$ and $r_2 = \lceil (\delta-1)(q-1)/q \rceil m$. Asymptotically the redundancy

$$r(B) \sim \lceil (\delta-2)(q-1)/q \rceil \log_q n$$

yields the estimate $\rho(q,n,d) \lesssim \lceil (d-2)(q-1)/q \rceil \log_q n$, since the inequality $d \geqslant \delta$ is valid for all BCH codes. Thus, for $q = 2$ (binary BCH codes), or $d = 3$, or $q = 3$ and $d = 5$ we have $\rho(q,n,d) \sim h(q,n,d)$. Other BCH codes can asymptotically meet the Hamming bound $h(q,n,d)$ if $d \geqslant 2\lceil (\delta-2)(q-1)/q \rceil + 1$, which implies that d is larger than the designed distance δ. The BCH codes that asymptotically satisfy the strict inequality $d > \delta$, are unknown. Thus two problems arise:

P1. Are there q-ary codes with distance $d > 3$, whose cardinality asymptotically exceeds that of BCH codes?

P2. Are there q-ary codes with distance $d > 3$ that asymptotically meet the Hamming bound?

2. THE RESULTS

There is a positive answer to P1 for all $GF(q), q \geqslant 3$ and $d \leqslant 6$. Namely, linear codes with distances $d = 4, 5$, and 6 are known [2] that give the estimate

(1) $$\rho(q, n, 4) \lesssim 1.5 \log_q n,$$

(2) $$\rho(q, n, 5) \lesssim 2.4 \log_q n,$$

(3,) $$\rho(q, n, 6) \lesssim 3 \log_q n$$

while BCH codes with designed distances 4,5, and 6 give the estimates:

$$\rho(q, n, 4) \lesssim 2 \log_q n, \ \ \rho(q, n, 5) \lesssim 3 \log_q n, \ \ \rho(q, n, 6) \lesssim 4 \log_q n.$$

The problem P1 for $d \geqslant 7$ is open for all q.

Almost nothing is known about the problem P2. The constructions [3], [4] similar to the BCH construction give optimal parameters for $q = 4$ and $d = 5$:

$$\rho(4, n, 5) \sim 2 \log_4 n.$$

Double-error-correcting codes which give the estimate

(4) $$\rho(q, n, 5) \lesssim (7 \log_q n)/3$$

over a field $GF(q)$ of odd characteristic have been constructed in the recent paper [6], thus improving the estimate (2). Below in Section 3 we generalize the construction [2] for arbitrary d. In Section 4 we combine the constructions [2] and [6] and obtain the redundancy (4) for all q-ary alphabets.

3. THE MAIN CONSTRUCTION

We consider the codes over the Galois field $L = GF(q)$. Let $F = GF(q^s), s = 2, 3, \ldots$ be an extension of degree s of the field L. Let L^s denote the vector space over L of dimension s. Below the elements of L will be denoted by greek letters and the elements of the extension F by latin letters.

Let $X = \{x_1, \ldots, x_n\} \subset F$ be a subset of $n < q^s$ different nonzero elements and let

$$H(s, X, j_1, \ldots, j_l) = (x_i^j),\ i = 1, \ldots, n,\ j = j_1, \ldots, j_l$$

be a $l \times n$ matrix generated by the elements of X in the powers $j_1, \ldots, j_l$. We define a code V_d with distance d in several steps. At first, define the parity-check matrix H_d of the code V_d in the form

$$\text{(5)} \qquad H_d = H(s, X, q^m + 1, \ldots, q^{m+d-2} + 1), \quad s \geqslant d - 1.$$

Let g be a primitive element of F. Note that if $X = F \backslash 0$ is the maximum nonzero set, V_d is a cyclic code (with zeros g^{q^i+1}, $i = m, \ldots, m+d-2$) which we hereafter denote by $\widehat{V}_d$.

Otherwise, V_d is the restriction (punctured code) of $\widehat{V}_d$ to the set X of length $n = |X|$.

Let us first estimate the minimum redundancy in the class of codes V_d and then estimate its distance. In order to minimize the redundancy (over arbitrary set X), specify the parameters s and m

$$\text{(6)} \qquad s \equiv d (\text{mod } 2), \quad m = \lceil s/2 \rceil - t, \quad t = \lfloor (d-1)/2 \rfloor$$

According to (6), the zeros of the code $\widehat{V}_d$ form t cyclotomic classes of cardinality s and, in addition, for even s, one cyclotomic class of cardinality $s/2$. Thus

$$\text{(7)} \qquad r(V_d) \leqslant s(d-1)/2$$

with equality for the code $\widehat{V}_d$. The estimate (7) corresponds to the Hamming bound $h(q, n, d)$ for odd d, if V_d has distance d and X has the same exponential order as F. Still, these two conditions do not hold simultaneously in the following constructions (see theorems 1,4). Note that $\widehat{V}_d$-code defined by (5,6), is well-known for $q = 2, d = 5$ [1, Sect. 9.11] and V_d-codes were studied for arbitrary q and $d = 4, 5$, and 6 [2].

Consider the distance of the code V_d defined by (5). Let $v \in L^n$ be an arbitrary vector of weight w with the set $\xi(v) = \{\xi_1, \ldots, \xi_w\}$ of w nonzero symbols in positions $f_1, \ldots, f_w$. Denote the corresponding locator set of v by $X(v) = \{y_1, \ldots, y_w | \, y_i = x_{f_i}, i = 1, \ldots, w\}$. Let $\overline{X}(v)$ be the subspace in L^s over L, generated by the elements of $X(v)$. Let $r = \lfloor w/2 \rfloor$.

THEOREM 1. *The code V_d defined by (5), contains a codeword v of weight $w < d$ iff* $\dim(\overline{X}(v)) \leqslant r$.

Theorem 1 generalizes Theorems 1-3 in [2] that concern distances $d = 4, 5$, and 6 respectively, and is proved by similar techniques.

Corollary 1. *The code V_d defined by (5), has distance at least d iff the condition*

$$\dim(\overline{Y}) > r \tag{8}$$

holds for all subsets $Y \subset X$ of cardinality $|Y| = w$ and all $w = 2, \dots, d-1$.

From now on we use the set X of locators $x = (\tau_0, \dots, \tau_{s-1})$ with fixed coordinate

$$\tau_0 = 1. \tag{9}$$

The codes V_d have distance at least 4, as all locators in the set X are noncollinear, according to (9). Consider the codes V_d, $d = 5, \dots, 8$. The next theorem specifies for any codewords v of weight $w < d$ the relations between the set $\xi(v)$ of symbols and the set $X(v)$ of locators and allows us to improve the estimate (2) [2].

Theorem 2. *For $d = 5, 6$ any codeword $v \in V_d$ of weight $w < d$ satisfies the following conditions* C1 *and* D1.

C1. *All w locators of the set $X(v)$ lie on the straight line $L(a,b) = a + b\lambda$ in L^s with parameters $a, b \in L^s$, $a = \{1, \alpha_1, \dots, \alpha_{s-1}\}$, $b = \{0, \beta_1, \dots, \beta_{s-1}\} \neq 0$ and variable $\lambda \in L$, i.e.*

$$y_i = a + b\lambda_i,\ i = 1, \dots, w. \tag{10}$$

D1. *The symbols ξ_i and variables λ_i are connected by the equations*

$$F(j) = \sum_{i=1}^{w} \xi_i \lambda_i^j = 0,\ j = 0, 1, 2. \tag{11}$$

Conversely, any w points on the line $L(a,b)$ (10) specify the codewords of weight w, satisfying (11).

Theorem 3. *For $d = 7, 8$ any codeword $v \in V_d$ of weight $w < d$ satisfies either conditions* C1, D1 *or the following conditions* C2, D2.

C2. *All w locators of the set $X(v), w = 6, 7$, lie on the plane $P(a,b,c) = a + b\lambda + c\eta$ with the parameters $a, b, c \in L^s$, $a = (1, \alpha, \dots, \alpha_{s-1})$, $b = (0, \beta_1, \dots \beta_{s-1}) \neq 0$, $c = (0, \theta_1, \dots, \theta_{s-1}) \neq 0$, and variables $\lambda, \eta \in L$, i.e.*

$$y_i = a + b\lambda_i + c\eta_i,\ i = 1, \dots, w. \tag{12}$$

D2. *The symbols ξ_i and variables λ_i, η_i are connected by the equations*

$$F(j_1, j_2) = \sum_{i=1}^{w} \xi_i \lambda_i^{j_1} \eta_i^{j_2} = 0, \quad 0 \leqslant j_1, j_2 \leqslant 2,\ 0 \leqslant j_1 + j_2 \leqslant 2. \tag{13}$$

Conversely, any w points on the plane $P(a,b,c)$ (12) specify the codewords of weight w, satisfying (13).

The conditions C1, C2 follow immediately from Corollary 1, taking into account the restriction (9). The conditions D1, D2 are derived by substitution of the "locator" equalities (10), (12), using the method of Theorem 1. Note also that Theorems 2,3 can be generalized for an arbitrary distance d, using conditions similar to (10)–(13) with consequently increasing the number of variables, according to (8).

4. Two methods for constructing the codes V_5, V_6.

There are two ways of constructing the codes V_5, V_6, according to Theorem 2.

A1. Using Condition C1, construct a locator set $X \subset L^s$ that contains three or less points on any line $L(a,b) = a + b\lambda$ with $a = (1, \alpha_1, \ldots, \alpha_{s-1}), b = (0, \beta_1, \ldots, \beta_{s-1}) \neq 0$. We are interested in constructing a set X of maximum possible cardinality, as the relative redundancy $r(V_d)/\log_q n$ decreases when $|X| = n$ grows.

A2. Using both the conditions C1 and D1, add new parity checks in the codes V_5, V_6 such that equations (10) and (11) do not satisfy these parity checks simultaneously. In this case we are interested in adding the minimum number of parity checks.

I. Consider the method A1. Any locator $x = (1, \tau_1, \ldots, \tau_{s-1})$ on the line $a + b\lambda$ can be represented in coordinate form

$$\tau_j = \alpha_j + \lambda\beta_j, \; j = 1, \ldots, s-1. \tag{14}$$

At first, let the set $X \subset L^s$ be a variety

$$F_3(\tau_1, \ldots, \tau_{s-1}) = 0 \tag{15}$$

of degree $\leqslant 3$, i.e. F_3 is a function of $s-1$ variables of degree 3 or lower. Substituting the coordinates (14) of the points on the line in equation (15), we obtain the equation in the parameter λ of degree $\leqslant 3$. If at least one coefficient of this equation is nonzero, then the equation has at most three roots, i.e. the line (14) intersects the variety (15) in three or less points. Otherwise, the equation is identically zero and all q points of the line (14) belong to the variety (15). Thus the problem is restricted to constructing the cubic varieties (15) of maximum cardinality containing no entire straight lines (14). One construction is given below.

Consider the field $GF(q^m)$ with a basis $g_1, \ldots, g_m$ over L and the norm [5] $N_m(x) = x^{q^{m-1}+q^{m-2}+\cdots+1}$ of an arbitrary element $x = \sum_{i=1}^m \tau_i g_i$ of $GF(q^m)$. It is well known that $N_m(x) = N_m(\sum_{i=1}^m \tau_i g_i) \triangleq N_m(\tau_1, \ldots, \tau_m)$ is the homogeneous form of variables $\tau_1, \ldots, \tau_m$ of degree m with nonzero values in L for any $x \neq 0$.

Now define the set $X = \{(1, \tau_1, \ldots, \tau_{s-1})\}$ by specification of each sixth coordinate $\tau_{6i-5}, i = 1, \ldots, h$, $h = \lceil (s-1)/6 \rceil$, as a function of the next 5 coordinates in the following way

$$\tau_{6i-5} = N_3(\tau_{6i}, \tau_{6i-1}, \tau_{6i-2}) + N_2(\tau_{6i-3}, \tau_{6i-4}), \quad i = 1, \ldots, h, \tag{16}$$

where $\tau_j \triangleq 0$ for $j > s-1$. Thus the constructed set $X \subset L^s$ has the exponential order $\log_q |X| = s - 1 - h = \lfloor 5(s-1)/6 \rfloor$.

THEOREM 4 [2]. *The constructed set $X = \{(1, \tau_1, \ldots, \tau_{s-1})\} \subset L^s$ that satisfies the condition (16), contains 3 or less points on every straight line $L(a, b) = (\alpha_0, \ldots, \alpha_{s-1}) + \lambda(\beta_0, \ldots, \beta_{s-1})$ in L^s.*

The proof is based on the following arguments. Substitute coordinates (14) into h equations (16). For $i = 1$ we have the following cubic equation for parameter λ:

$$\begin{aligned} &N_3(\alpha_6 + \lambda\beta_6, \alpha_5 + \lambda\beta_5, \alpha_4 + \lambda\beta_4) + N_2(\alpha_3 + \lambda\beta_3, \alpha_2 + \lambda\beta_2) + \alpha_1 + \lambda\beta_1 \\ &\qquad \triangleq \mu_3\lambda^3 + \mu_2\lambda^2 + \mu_1\lambda + \mu_0 = 0. \end{aligned} \tag{17}$$

It can be easily checked that $\mu_3 = N_3(\beta_6, \beta_5, \beta_4)$ and is equal to zero iff $\beta_6 = \beta_5 = \beta_4 = 0$. In this case $N_3(\cdot, \cdot, \cdot)$-function does not depend on λ. Therefore $\mu_2 = N_2(\beta_3, \beta_2)$ and is equal to zero iff $\beta_3 = \beta_2 = 0$. In this case $\mu_1 = 0$ iff $\beta_1 = 0$. Thus the equation (17) becomes an equality only if $\beta_1 = \beta_2 = \cdots = \beta_6 = 0$. Similarly, all h equations (16) can become equalities only if $\beta_1 = \cdots = \beta_{s-1} = 0$, which is impossible for the straight line $a + b\lambda$ with $b = (0, \beta_1, \ldots, \beta_{s-1}) \neq 0$. Therefore at least one cubic equation is nontrivial and gives at most 3 possible values for variable λ.

From Theorems 2 and 4 we obtain

COROLLARY 2. *Codes V_5, V_6, defined in equations (5) and (6) with the locator set X defined by the system (16) give the asymptotical estimates (2),(3).*

II. Now consider the following construction that satisfies the method A2. Below $X = \{x_1, \ldots, x_n\}$ is the complete set of $n = q^{s-1}$ elements of L^s that satisfy (9). For any element $x = (1, \tau_1, \ldots, \tau_{s-1}) \in X$ define the vector $f_m(x) = (f_{m,1}, \ldots, f_{m,l})$ over L of length $l = \lceil (s-1)/m \rceil$, where

$$f_{m,j} = f_{m,j}(x) = N_m(\tau_{mj+1}, \ldots, \tau_{(m+1)j}), j = 0, 1, \ldots, l-1, \tag{18}$$

and $\tau_k \triangleq 0$ for $k > s-1$. Consider the matrix $f_m = (f_m(x_i)^T, i = 1, \ldots, n)$ of size $l \times n$ and the code F_m with parity check matrix f_m. Finally, define the subcodes V_5' and V_6' of codes V_5 and V_6 (see (5),(6)) as:

$$V_5' = V_5 \bigcap F_3, V_6' = V_6 \bigcap F_3 \bigcap F_4. \tag{19}$$

THEOREM 5. *The code V_5' has the parameters*

$$n = q^{s-1}, r = 2s + \lceil (s-1)/3 \rceil, d \geqslant 5, s = 5, 7, \ldots \quad (20)$$

The proof is based on the following arguments. Suppose there is a codeword $v \in V_5'$ of weight $w = 4$. The locators y_i and symbols $\xi_i, i = 1, \ldots, w$, of the codeword v satisfy equations (10) and (11). Besides, the codeword v satisfies the following l parity check equations over the matrix f_3:

$$\sum_{i=1}^{w} \xi_i f_{3,j}(y_i) = 0, \; j = 0, 1, \ldots, l-1.$$

It can be shown that the substitution of linear equations (10) for symbols $f_{3,j}(y_i)$ in equations (18) gives the equation $\sum_{i=1}^{w} \xi_i \lambda_i^3 = 0$. Therefore we have for ξ_i a nondegenerate system of equations: $\sum_{i=1}^{4} \xi_i \lambda_i^k = 0, k = 0, \ldots, 3$, with the unique solution $\xi_i = 0, \; i = 1, \ldots, 4$.

The parameters (20) give estimates (4), thus improving (2) for all q.

THEOREM 6. *The code V_6' has the parameters*

$$n = q^{s-1}, r = 5s/2 + \lceil (s-1)/3 \rceil + \lceil (s-1)/4 \rceil, \; d \geqslant 6, \; s = 6, 8, \ldots. \quad (21)$$

The proof is similar to the proof of Theorem 5 and gives the system $\sum_{i=1}^{5} \xi_i \lambda_i^j = 0, \; j = 0, \ldots, 4$, for the symbols ξ_i.

Note that parameters (21) are slightly worse than parameters obtained from (3).

5. CONCLUDING REMARKS

Theorems 2,4, and 5 hold also for other constructions. Consider the code U_5 over $L = GF(2^m)$ with parity check matrix

$$H(s, X, 1, (q^s + q)/2), \; s = 2, 3, 4, \ldots, \quad (22)$$

and the set $X = \{1, \tau_1, \ldots, \tau_{s-1}\}$. It can be shown that condition C1 (10) is also valid for the code U_5 [2] and condition D1 (11) holds in a slightly different form:

$$\sum_{i=1}^{w} \xi_i^2 \lambda_i^j = 0, \; j = 0, 1, 2.$$

Therefore specification of the set X according to (16) gives $d \geqslant 5$ also for the code U_5. Using the elements $h_{m,j} = \sqrt{f_{m,j}}$ in the parity check matrix of the code F_3 (see (18),(19)) instead of elements $f_{m,j}$, we obtain Theorem 5 for the code $U_5' = U_5 \bigcap F_3$. The other construction $W_5' = W_5 \bigcap F_3$, similar to the codes V_5', U_5', is obtained [6] for the fields L of odd characteristic by the code W_5 with the parity check matrix

$$H(s, X, 0, 1, 2), \ s = 1, 2, \ldots . \tag{23}$$

Note that constructions U_5' (22) and W_5' (23) have a simple decoding procedure and besides, W_5' admits a cyclic representation. These codes improve the BCH codes starting from the length $n = q^3$.

Finally, note that the restriction of q-ary codes to the alphabet of size $q' < q$ [7] expands the asymptotic estimates (1)-(4) for any alphabet.

REFERENCES

1. F.J.MacWilliams and N.J.A.Sloane. *The Theory of Error-Correcting Codes.* North-Holland, Amsterdam, 1976.

2. I.I.Dumer. Nonbinary codes with distances 4,5, and 6 of cardinality greater than the BCH codes, *Problemy Peredachi Inform.*, vol. 24, N 3 (1988), 42–54.

3. D.N.Gevorkyan, A.M.Avetisyan, and V.A.Tigranyan. On the construction of codes correcting two errors in Hamming metric over Galois fields. In *Vychislitel'naya Technika*, Kuibyshev, N 3 (1975), 19–21 (in Russian).

4. I.I.Dumer and V.A.Zinoviev. Some new codes over Galois fields GF(2^r), In *Proc. 4th Intern. Sympos. Inform. Theory*, Pt.2, Moscow-Leningrad (1976), 37–39 (in Russian).

5. Z.I.Borevich and I.R.Shafarevich. *Number Theory.* Moscow, Nauka, 1972 (in Russian).

6. I.I.Dumer. Nonbinary codes with distance 5. In *Proc. 2nd Intern. Workshop on Algebraic and Combinatorial Coding Theory*, Leningrad, Moscow (1990), 63–66.

7. D.N.Gevorkyan. On nonbinary codes with fixed code distance. In *Proc. 5th Intern. Symp. Inform. Theory*, Abstracts of papers, Pt.1 , Moscow-Tbilisi (1979), 93–96 (in Russian).

Saddle Point Techniques in Asymptotic Coding Theory

Danièle GARDY *
L.R.I., CNRS-URA 410,
Bât. 490, Université Paris XI,
91405 Orsay, France
&
Patrick SOLÉ †
Laboratoire I.3.S., CNRS-URA 1376,
Bât. 4, 250 rue Albert Einstein,
Sophia-Antipolis, 06560 Valbonne, France.

Abstract

We use asymptotic estimates on coefficients of generating functions to derive anew the asymptotic behaviour of the volume of Hamming spheres and Lee spheres for small alphabets. We then derive the asymptotic volume of Lee spheres for large alphabets, and an asymptotic relation between the covering radius and the dual distance of binary codes.

1 Introduction.

From a graph theoretic point of view, codes for the Hamming metric are sets of vertices in the n-dimensional hypercube. Due to the cartesian product structure of this graph (or of the n-dimensional torus, which is the graph adapted to the Lee metric) many statistics of interest (surface of spheres, for instance) are additive; this leads to generating functions which are n^{th} powers of the generating function for the one-dimensional case.

This type of generating function has been extensively studied in statistics [4, 7], and in analysis of algorithms [6, 8]. The techniques used to get an asymptotic approximation of their coefficients involve complex analysis [3, 5], and in particular the method of steepest descent to estimate contour integrals.

The paper is organized as follows. We recall the analytic results that we need in Section 2. We rederive the classical estimate for the volume of Hamming spheres via the entropy function in Section 3, and give in Section 4 a simpler proof of an estimate of Astola [1] for the volume of the Lee spheres, for small alphabets of odd cardinality. We also derive a similar result when this cardinality is even. We present in Section 5 a new estimate for Lee spheres over large alphabets, which is expressed using a function not essentially more complicated than the binary entropy function. Finally we derive in Section 6 an asymptotic relation between the covering radius of a binary code and its dual distance.

2 The Saddle Point Method

In this section our aim is to estimate the coefficient of order r of a generating function $\Phi(z)$, a quantity that we shall denote henceforth by $[z^r]\{\Phi(z)\}$. More precisely, we are interested in generating functions of the kind $\Phi(z) = f(z)^n g(z)$, and in their coefficients $[z^r]\{f(z)^n g(z)\}$ for large n and $r \sim \lambda n$.

In order to use complex variable techniques, we need to introduce Cauchy's formula [3, p.72].

*This work was partially performed while this author was visiting Brown University, Providence, R.I., U.S.A. The author also acknowledges the support of the PRC Mathématique-Informatique (CNRS) and of ESPRIT-II Basic Research Action No. 3075 (project ALCOM).

†This work was partially supported by the PRC C^3 (CNRS and MRT).

Lemma 1 *Let D denote a simply connected domain containing the origin and where Φ is analytic. Let Γ denote a simply closed contour contained in D. Then*

$$[z^r]\{\Phi(z)\} = \frac{1}{2\pi i}\int_\Gamma \Phi(z)\frac{dz}{z^{r+1}}.$$

Rewriting the integrand in Cauchy's formula in the form $e^{h(z)}$ with $h(z) = \log\Phi(z) - (r+1)\log(z)$, we are left to estimate a contour integral $I_{r,n}$, say, of the type

$$I_{r,n} = \frac{1}{2\pi i}\int_\Gamma e^{h(z)}dz.$$

The basic idea of the saddle point method is to use a second order Taylor approximation of $h(z)$ about a point ρ where $h'(\rho) = 0$. This point is called a *saddle point* because in its neighborhood the surface $z \mapsto h(z)$ resembles a saddle. The part of Γ nearby the saddle point is the most important contribution in $I_{r,n}$ and we obtain

$$I_{r,n} \approx \frac{e^{h(\rho)}}{\sqrt{2\pi h''(\rho)}}.$$

This holds under suitable conditions on ρ, e.g. $h''(\rho) > 0$. To get a condition bearing on r only, we define two operators acting on a function f

$$\Delta f(z) = zf'(z)/f(z) \quad \text{and} \quad \delta f(z) = (\Delta f)'(z)/z.$$

It can be shown that, when f is a power series with positive coefficients, then $\delta f(\rho) > 0$ (see for example [6, p.65]). We shall say that f is *degenerate* if $f(0) = 0$ or if there exists an analytic function h and an integer m such that $f(z) = h(z^m)$. We summarize the discussion in the following statement, which is essentially due to [4] but is best given in the version of [7, p.868]:

Theorem 1 *Let f be a power series with real positive coefficients and non degenerate; assume that the equation $\Delta f(z) = (r+1)/n$ has a real positive solution ρ. If f is analytic in an open set including the disk of radius ρ and center the origin, then for large n and r, and with r/n restricted to an interval $[A, B]$ $(A > 0)$*

$$[z^r]\{f(z)^n\} = \frac{f(\rho)^n}{\rho^{r+1}\sqrt{2\pi n\delta f(\rho)}}(1 + o(1)).$$

This result is easily extended to take into account a factor $g(z)$, as long as g does not introduce singularities closer to the origin than the saddle point:

Theorem 2 *Let f and g be non degenerate power series with positive coefficients. Assume that the equation $\Delta f(z)+\Delta g(z)/n = (r+1)/n$ has a real positive solution ρ. If f and g are analytic in an open disk including the circle of radius ρ and center the origin, then for large n and r with r/n restricted to an interval $[A, B]$ $(A > 0)$*

$$[z^r]\{f(z)^n g(z)\} = \frac{f(\rho)^n g(\rho)}{\rho^{r+1}\sqrt{2\pi n\delta f(\rho)}}(1 + o(1)).$$

Proof: The proof of Theorem 2 is similar to that of Theorem 1 and to a method of Hayman [9][13, Ch.5]. For this reason we only give a sketch of the proof.

To approximate the integral $\int e^{h(z)}dz$, with $h(z) = n\log f(z) + \log g(z) - (r+1)\log z$, we first compute the saddle point ρ (or an approximate value), which is defined by $h'(z) = 0$. We then choose as integration path the circle of center 0 and radius ρ, and divide the integral in two parts. The first part comprises the values of $z = \rho e^{i\theta}$ which are close to ρ: $|\theta| < \alpha$, for some suitable small α. For these values, the function $h(\rho e^{i\theta})$ has a second order Taylor expansion with an error term $O(\theta^3)$; we plug this expansion into the integral, which we then extend to a gaussian integral of known value.

We next have to show that the part of the integral far from the saddle point $(\alpha \le |\theta| \le \pi)$ is negligeable. But this follows from the fact that the function $|g(z)|$ is maximal on the real axis, that the function $|f(z)|$ is maximal on the real axis and nowhere else on the circle $\{z = \rho e^{i\theta}\}$, and that $|f(z)^n g(z)|$ decreases exponentially when n increases. □

In general, Theorems 1 and 2 are difficult to apply, because ρ is a function of n and r, and $f(\rho)^n$ is difficult to estimate. So, we shall use the following corollary.

Corollary 1 *Under the hypotheses of Theorem 2, with $r/n = \lambda$ and with $\Phi(z) = f(z)^n g(z)$,*

$$\frac{1}{n}\log([z^r]\{\Phi(z)\}) = \log(f(\rho)) - \lambda\log(\rho) + o(1).$$

We can also extend Corollary 1 to use a simpler saddle point:

Corollary 2 *Corollary 1 is still valid if we use the simpler saddle point defined by $\Delta f(z) = r/n$.*

Proof: The points ρ, defined by $\Delta f(z) + \Delta g(z)/n = (r+1)/n$, and ρ_0, defined by $\Delta f(z) = r/n$, are such that $\rho_0 = \rho(1 + O(1/n))$. Hence $\log f(\rho_0) - \lambda\log(\rho_0) = \log f(\rho) - \lambda\log(\rho) + o(1)$. □

3 Hamming Spheres

Recall that the Hamming weight $W(\mathbf{x})$ of $\mathbf{x} = (x_1, \ldots, x_n) \in \mathbf{F}_q^n$ is $W(\mathbf{x}) = Card\{i \in [1, \ldots, n] | x_i \neq 0\}$. The Hamming sphere of radius r centered at $(0, \ldots, 0)$ is $\mathbf{B}_r = \{\mathbf{x} \in \mathbf{F}_q^n | W(\mathbf{x}) \leq r\}$; its volume is $|\mathbf{B}_r|$. The generating function for the volume of the Hamming spheres is $\Phi_n(z) = \sum_{r=0}^{n} |\mathbf{B}_r| z^r = (1+z)^n/(1-z)$. Let $f(z) = 1 + z$ and $g(z) = 1/(1-z)$; then $\Phi_n(z) = f(z)^n g(z)$. The function f is entire, with positive coefficients, and non-degenerate, and the function g has positive coefficients and a simple pole in 1. The following result can be derived by more elementary but also more tedious means.

Theorem 3 *Let $H(x) = -x\log_2(x) - (1-x)\log_2(1-x)$ be defined on $(0,1)$. If $\lambda < 0.5$, then for $r \sim \lambda n$ and large n*

$$\frac{1}{n}\log_2([z^r]\{\Phi_n(z)\}) = H(\lambda) + o(1).$$

Proof: To check that the function $\Phi_n(z)$ satisfies the assumptions of Theorem 2, we have to compare the singularity of g, which is 1, to the saddle point defined by the equation

$$\frac{z}{1+z} + \frac{z}{n(1-z)} = \frac{r+1}{n}.$$

This has for solution $\rho = \lambda/(1-\lambda)$, which is < 1 iff $\lambda < 1/2$. The result follows by Corollary 2. □

We leave as an exercise to the reader to derive an analogous result for q-ary codes (Cf. [12] Lemma 5.1.6 p.55).

4 Lee Spheres for Small Alphabets

Let $s = \lfloor q/2 \rfloor$ and $\mathbf{Z}_q = \{-s, -(s-1), \ldots, 0, \ldots, (s-1), s\}$. We recall that the Lee weight of $x \in \mathbf{Z}_q$ is

$$\begin{aligned} W_L(x) &= x \quad \text{if } x \geq 0; \\ &= -x \quad \text{if } x < 0. \end{aligned}$$

The Lee weight of $\mathbf{x} = (x_1, \ldots, x_n) \in \mathbf{Z}_q^n$ is $W_L(\mathbf{x}) = \sum_{i=1}^{n} W_L(x_i)$. The Lee sphere of radius r is $\mathbf{B}_r^L = \{\mathbf{x} \in \mathbf{F}_q^n | W_L(\mathbf{x}) \leq r\}$; its volume is $|\mathbf{B}_r^L|$. Then we know from [2, p.298] that the generating function $\Phi_{n,q}(z) = \sum_{r \geq 0} |\mathbf{B}_r^L| z^r$ can be evaluated as

$$\Phi_{n,q}(z) = \frac{f(z)^n}{1-z}$$

with

$$\begin{aligned} f(z) &= 1 + 2\sum_{i=1}^{s} z^i \quad (q = 2s+1); \\ f(z) &= 1 + 2\sum_{i=1}^{s-1} z^i + z^s \quad (q = 2s). \end{aligned}$$

Let

$$L(\tau, q) = \lim_{n \mapsto +\infty} (1 - \frac{1}{n}\log_q([z^{ns\tau}]\{\Phi_{n,q}(z)\})).$$

The following result was proved by Astola [1] using multinomial coefficients and Lagrange multipliers.

Theorem 4 *Let $q = 2s+1$. Then $L(\tau, q) = 1 + \log_q(\alpha\rho^{\tau s})$, where $\alpha \geq 0$ and $\rho \geq 0$ are defined by the two equations*

$$\alpha(1 + 2\sum_{i=1}^{s} \rho^i) = 1 \qquad \text{and} \qquad \alpha \sum_{i=1}^{s} i\rho^i = \frac{\tau s}{2},$$

and where $0 \leq \tau \leq (q+1)/(2q)$. Moreover $L(\tau, q) = 0$ if $\tau \geq (q+1)/(2q)$.

Our method allows us to get a new derivation of Astola's result (Theorem 5) when q is odd and to prove a similar result (Theorem 6) when q is even. The term ρ in Theorems 4 and 5 is the same (it is defined by the same equation), and we have the relation $\alpha = 1/f(\rho)$.

Theorem 5 *Let $q = 2s+1$. If $\tau < (q+1)/(2q)$, then*

$$\frac{1}{n}\log_q([z^{ns\tau}]\{\Phi_{n,q}(z)\}) = \log_q(f(\rho)) - s\tau\log_q(\rho) + o(1),$$

where ρ is the unique real positive solution of

$$2\sum_{i=1}^{s}(i - s\tau)z^i = s\tau, \tag{1}$$

and where $f(\rho) = (s\rho + s + 1)/(s(\rho\tau - \rho - \tau) + s + 1)$.

Proof: Let $\sigma(z) = \sum_{i=1}^{s} z^i$. Then, we see that $f(z) = 1 + 2\sigma(z)$. Tedious but straightforward calculations show that $\sigma(z)$ satisfies the first order ODE

$$z(z-1)\sigma'(z) + (-sz + s + 1)\sigma(z) = sz.$$

The saddle point ρ satisfies the equation

$$2z\sigma'(z) = s\tau(1 + 2\sigma(z)).$$

Getting rid of $\sigma'(z)$ between these two equations yields an expression for $\sigma(z)$, hence the above-mentioned expression for $f(\rho)$. The condition on τ comes from $\Delta f(1) > s\tau$, and ensures that the saddle point $\rho = (\Delta f)^{-1}(s\tau)$ is smaller than 1, the pole of g. □

Theorem 6 *Let $q = 2s$. If $\tau < 1/2$, then*

$$\frac{1}{n}\log_q([z^{ns\tau}]\{\Phi_{n,q}(z)\}) = \log_q(f(\rho)) - s\tau\log_q(\rho) + o(1),$$

where ρ is the unique real positive solution of

$$2\sum_{i=1}^{s}(i - s\tau)z^i = s\tau + s(1-\tau)z^s, \tag{2}$$

and where $f(\rho) = (s\rho + s + 1 - \rho^s)/(s(\rho\tau - \rho - \tau) + s + 1)$.

Proof: Here $f(z) = 1 + 2\sigma(z) - z^s$, with $\sigma(z) = \sum_{i=1}^{s} z^i$ as above. The saddle point satisfies the equation

$$2z\sigma'(z) - sz^s = s\tau(1 + 2\sigma(z) - z^s).$$

As before, we get rid of $\sigma'(z)$ and get an expression for $\sigma(z)$, which gives readily the expression for $f(\rho)$. The bound on τ comes again from $\Delta f(1) > s\tau$. □

5 Lee Spheres for Large Alphabets

We note that, for $q \geq 2r + 1$:

$$[z^r]\{\Phi_{n,q}(z)\} = [z^r]\{\frac{(1 + 2\sum_{i\geq 1} z^i)^n}{1-z}\} = [z^r]\{\frac{(1+z)^n}{(1-z)^{n+1}}\}.$$

The following result appears to be new or at least unpublished.

Theorem 7 *Let $V_{n,r}$ denotes the volume of the Lee sphere of radius r in Z_q^n. When $q \geq 2r+1$ this quantity does not depend on q, and for large n and r/n going to λ ($\lambda > 0$), we get*

$$\frac{1}{n}\log_q(V_{n,r}) = L_q(\lambda) + o(1),$$

where

$$L_q(x) = x\log_q(x) + \log_q(x + \sqrt{x^2+1}) - x\log_q(\sqrt{x^2+1} - 1).$$

Proof: We have $V_{n,r} = [z^n]\{(1+z)^n/(1-z)^{n+1}\}$. The saddle point equation can be written as

$$(r+2)z^2 + (2n+1)z - (r+1) = 0.$$

The only positive root is $\rho = (\sqrt{(2n+1)^2 + 4(r+1)(r+2)} - (2n+1))/(2(r+2))$. When n is large and r/n goes to λ this is

$$\rho \sim \frac{\sqrt{1+\lambda^2} - 1}{\lambda}.$$

It is easily checked that this quantity is always < 1. Hence the pole of g is not a problem and Corollary 1 gives the result. □

Using this result we obtain the analogues of the Hamming bound [2, p.299] and the Gilbert-Varshamov bound [2, p.321]:

Corollary 3 *Let $R(\lambda)$ denote the largest achievable rate of a family of codes of minimum Lee distance λn for large n, and such that $q \geq 2\lambda n + 1$. Then, for $0 \leq \lambda < q/e$ we have*

$$1 - L_q(\lambda) \leq R(\lambda) \leq 1 - L_q(\frac{\lambda}{2}).$$

Proof: We have the result for $\lambda \leq 2L_q^{-1}(1)$. But $L_q(x) = \log_q(2ex) + O(1/x^2)$ for large x; hence the solution of the equation $L_q(x) = 1$ for large q is $x = q/(2e)(1+o(1))$.

Open Problem 1 *Are there families of codes which are better than the lower-bound?*

Open Problem 2 *Study $V_{n,r}$ when n and r are both large but no longer proportional.*

6 Covering Radius and Dual Distance

A current research problem in Coding Theory is to find upper bounds on the covering radius as a function of the dual distance. A connection with zeroes of Krawtchouk polynomials was discovered in [11]. A simpler power-sum approach was initiated in [10]. Here we derive an asymptotic version of Theorem 6 of [10], which we recall below:

Theorem 8 *Let C be a code of dual distance at least $2s+1$. Then its covering radius ρ is bounded from above by*

$$\rho \leq \frac{n}{2} - (2^{s/(s+1)} - 1)\mu_s(n)^{1/2s},$$

where $\mu_s(n) = [t^s/(2s)!]\cosh^n(\sqrt{t}/2)$.

To meet this goal we need the following Lemma.

Lemma 2 *Assume that the ratio $2s/n$ goes to a constant $\lambda \in]0,1]$ for n and s large. Then $(1/n)\,\mu_s(n)^{1/2s}$ goes to $\lambda\cosh(x_0)^{1/\lambda}/(2ex_0)$ where x_0 is the positive solution of $x\tanh(x) = 2\lambda$.*

Proof: Here $f(z) = \cosh(\sqrt{z}/2)$ is entire with positive coefficients and non degenerate, so we apply Theorem 1. Letting $x = \sqrt{z}/2$, the saddle point equation can be written as

$$x\tanh(x) = 2\lambda.$$

λ	Our bound	Tietäväinen's bound
.5	.249	0
.4	.276	0.010
.3	.306	0.042
.2	.342	0.100
.1	.388	0.200
.01	.465	0.400
.005	.475	0.429

Figure 1: Bounds for the covering radius ($\rho_0 = \lim \rho/n$)

Let x_0 be the unique real positive solution of this equation. Corollary 1 applied to the saddle point $z_0 = \sqrt{x_0}/2$ gives

$$\left(\frac{\mu_s(n)}{(2s)!}\right)^{1/2s} \sim \frac{\cosh^{1/\lambda}(x_0)}{2x_0}.$$

Stirling's approximation yields easily

$$((2s)!)^{1/2s} \sim \frac{2s}{e}.$$

□

Theorem 9 *Let C be a code of length n, covering radius ρ, and dual distance at least $2s+1$. Assume n large, ρ/n having ρ_0 as a limit, and $s \sim \lambda n$. Then, we have*

$$\rho_0 \leq \frac{1}{2} - \frac{\lambda}{2e} \frac{\cosh(x_0)^{1/\lambda}}{x_0}$$

where x_0 is the unique positive solution of $x \tanh(x) = 2\lambda$.

Proof: Dividing up the bound of Theorem 8 by n, and using Lemma 2 yields the desired result. □

Numerical computations show that this asymptotic bound is less precise than the bound $\rho_0 \leq 0.5 - \sqrt{\lambda(1-\lambda)}$ obtained in [11], although the bounds are in closer agreement for small λ. We give some results in Figure 1.

Open Problem 3 *Develop analogous results for q-ary codes.*

7 Conclusion

We hope to have demonstrated in this article the wide range of applicability of the saddle point approximation in asymptotic problems of Information Theory and Combinatorial Coding Theory. Many questions remain open. In particular many families of Lee codes have distance growing faster than n, and the asymptotic problems of Sections 4 and 5 are well worth studying for that case. Theorems 1 and 2 are no longer valid as stated when n and r are both large but their ratio does not have a constant, non null, limit; any extension of the asymptotic results of this paper to such cases thus requires a similar extension of Theorems 1 and 2. We hope to present this in a forthcoming paper.

References

[1] J. Astola, "On the Asymptotic Behaviour of Lee codes," Discr. Appl. Math, Vol. 8, pp. 13-23 (1984).

[2] E.R. Berlekamp, *Algebraic Coding Theory*, Aegean Park Press (1984).

[3] H. Cartan, *Théorie élémentaire des fonctions analytiques d'une ou plusieurs variables complexes*, Hermann (1961).

[4] H.E. Daniels, "Saddlepoint Approximation in Statistics," Ann. Math. Stat., Vol. 25, pp. 631-650 (1954).

[5] P. Henrici, *Applied and Computational Analysis*, Wiley (1977).

[6] D. Gardy, *Bases de données, allocations aléatoires: quelques analyses de performances*, Thèse d'Etat, Université Paris-Sud, Orsay (1989).

[7] I.J. Good, "Saddle point methods for the multinomial distribution," Ann. Math. Stat., Vol. 28, pp. 861-881 (1957).

[8] D.H. Greene, D.E. Knuth, *Mathematics for the analysis of algorithms*, Birkhäuser Verlag (1982).

[9] W.K. Hayman, "A generalisation of Stirling's formula," Journal für die reine und angewandte Mathematik, Vol. 196, pp. 67-95 (1956).

[10] P. Solé, K. G. Mehrothra, "A Generalization of the Norse Bound to Codes of Higher Strength," IEEE Trans. Information Theory, IT-37, pp. 190-192 (1991).

[11] A. Tietäväinen, "An Upper Bound on the Covering Radius as a Function of the Dual Distance," IEEE Trans. Information Theory, IT-36, pp. 1472-1474 (1990).

[12] J. H. van Lint, *Introduction to Coding Theory*, Springer, Graduate Texts in Math. 86 (1982).

[13] H.S. Wilf, *Generatingfunctionology*, Academic Press (1990).

NON-BINARY LOW RATE CONVOLUTIONAL CODES WITH ALMOST OPTIMUM WEIGHT SPECTRUM

Sergei I. Kovalev
Leningrad Aircraft Equipment Institute
Hertsen str. 67, 190000 Leningrad USSR

Abstract. We propose a construction of convolutional codes over the alphabet of size 2^k with code rate of one bit per symbol and constraint length $3k$, $k=2,3,\ldots$. These codes have the least possible values of several first elements of the weight spectrum for large values of k.

1. INTRODUCTION.

The transfer function $T(D)$ or the weight spectrum is an important feature of convolutional codes. The transfer function is used for the computation of the union bound on decoding error probability.

The well-known constructions of non-binary convolutional codes fall into two groups. The first group includes codes with q-ary input and output alphabets (for example [1-3]). Let d_f be the minimum weight of codewords and A_{d_f} be the number of codewords of weight d_f. For any code of this group we have $A_{d_f} \geqslant q-1$, because the product of a codeword and of any nonzero element of the field belongs to the code. For codes of the second group this estimate can be improved .

Codes of the second group have p-ary input and p^k-ary output alphabets. For example, consider orthogonal codes described by Viterbi [4]. The encoder of these codes contains a binary shift register of length k. Every information bit is put into the encoder during a time unit and the state of the register is transformed into a 2^k-ary output symbol. The code rate of this code is equal to one bit per symbol. Below we consider only codes of the second group.

In general, the encoder of a 2^k-ary time-invariant convolutional code of rate one bit per symbol contains a binary shift register of length L. Some of the delay elements are connected to adders modulo 2. Each information bit corresponds to a 2^k-ary symbol formed at the outputs of the adders. Evidently, in this case we have $d_f \leqslant L$.

In this paper we propose a construction of convolutional codes with $L = 3k$. These codes have about k first non-zero spectrum elements of least possible values. The following statements concern the case $L = 3k$, though some of them may be generalized.

2. MAIN RESULTS

Denote by I_l a binary sequence of length l . The length of the sequence is the number of symbols between its first and last nonzero positions including these positions themselves.

THEOREM 1. Let A_i, $i=d_f, d_f+1$, ... be the weight spectrum of a convolutional code. Then

$$\sum_{i=d_f}^{L+l-1} A_i \geqslant 1 + \sum_{j=2}^{l} 2^{(j-2)} , \; 0 < l \leqslant L .$$

Proof. The right hand side of this expression is equal to the number of different nonzero sequences I_l ,$0< l \leqslant L$. Every such sequence stays in the encoder during $L+l-1$ time units. Hence, all of these sequences correspond to codewords of weight not greater than $L+l-1$. ∎

A code is called s-optimal ($s \leqslant L$) if the elements of its spectrum are equal to $A_{d_f=L}= 1$, $A_{d_f+i} =2^{(i-1)}$, $i=1,2, \ldots ,s-1$. In other words, d_f is as large as possible and the weight spectrum satisfies Theorem 1 with equality for all $l = 1,\ldots,s$. Suppose a sequence I_l corresponds to a codeword of weight $L+l-1-\Delta$. The parameter Δ is called the defect of a sequence or of a codeword. It is easy to see that a code is s-optimal iff all sequences I_l with $l \leqslant s+1$ have zero defect, and all sequences I_l with $l=s+1+\delta$, $\delta \geqslant 0$ have the defect $\Delta \leqslant \delta$.

THEOREM 2. For an s-optimal code, $s < k$.

Proof. Let $g_i(x)$, $i=\overline{1,k}$ be generating polynomials determined by the connections of the encoder register with adders modulo 2. The binary representations of these polynomials form the matrix G of size $k \times L$, such that $aG^T= b$, where a is the content of the encoder register and b is the binary representation of the output symbol. The rank of the matrix G is not greater than k. Hence, any $k+1$ columns of G are linearly dependent. Therefore there exists a sequence I_l $(l \leqslant k+1)$ with nonzero defect . ∎

Consider generating polynomials of the code $g_i(x) = x^i + x^{2k-i-1} + x^{2k+i}$, $i=0,1,\ldots,k-1$. The shift register of the encoder can be separated into three parts $|A|B|C|$. Denote by X (resp. Y) the first (resp. the last) one of I_l. Suppose the encoding starts when X enters the register and ends when Y leaves it. The encoding process contains the following stages:

$$1)\ X \in A,\ 2)\ X \in B,\ 3)\ Y \in B,\ 4)\ Y \in C. \qquad (1)$$

The second and third stages may overlap. Note that every adder of the encoder has only one connection with each of the registers A, B and C. Hence, $2k$ nonzero output symbols are formed at the first and fourth stages. Let us consider the second stage. The encoder is shown in Fig.1. (Register C contains zeros).

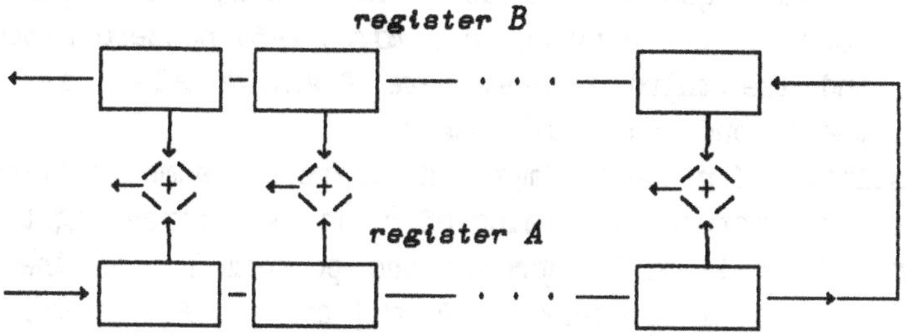

Fig.1. Encoder for stage 2.

Symbol X goes through the register B from right to left and information symbols in register A go from left to right.

LEMMA 1. Suppose an output symbol of the encoder is equal to zero and register A contains the sequence $0^z 1$, where 0^z means the sequence of z zeros. Then the next output symbol will be equal to zero after $\lceil Z/2 \rceil + 1$ or more time units.

Proof. An output symbol of the encoder is equal to zero when the contents of registers A and B coincide. Therefore the register B contains the sequence $0^z 1$ (see Fig.1). This 1 moves to the left in register B and the zero sequence 0^z moves to the right in the register A. The sum of these sequences will be nonzero during $\lceil Z/2 \rceil$ time units. ■

Note. Lemma 1 is valid even if the link between registers A and B is broken and an arbitrary sequence enters register B.

THEOREM 3. The number of zero output symbols formed at the second (or third) stage of the encoding is not greater than three.

Proof. Suppose the first nonzero output symbol appears at the time unit T_1 of the second stage of the encoding . At this moment the last $k-T_1$ delay elements of register B contain zeros. Hence the first $k-T_1$ delay elements of the register A are zero. The nonzero symbol in register A and the symbol X in register B enter the same adder modulo 2. According to Lemma 1 the next zero at the encoder output appears at the time moment $T_2 \geqslant T_1 + \lceil (k-T_1)/2 \rceil + 1$. After the encoder has formed the second zero output symbol, the first delay elements of the register

A contain the sequence $0^{k-T_2}1$, $k-T_2 \leqslant k/2$. This sequence stays in register A during at least $k/2$ time units, i.e. until the end of stage 2. Thus, $T_3 \geqslant T_2 + \lceil (k-T_2)/2 \rceil + 1$, $T_4 \geqslant T_3 + \lceil (k-T_2)/2 \rceil + 1 > k$. Hence T_4 does not fall into stage 2. The statement of theorem for stage 3 follows from the equivalence of the encoders $|A|B|$ and $|B|C|$.■

We see that after the encoding of the sequence I_l the corresponding code sequence contains $2k$ nonzero symbols after the first and fourth stages of encoding and $r \geqslant max(0,T-6)$ nonzero symbols after the second and the third stages. Here $T=min(2k,k+l-1)$ is the total duration of the second and third stages.

The estimate of r can be improved. There are some modifications of the encoder that increase the value of r. It is interesting to consider the following method which improves the performance of the code. We consider two identical encoders $|A_1|B_1|C_1|$ and $|A_2|B_2|C_2|$ and construct the combined encoder $|A_1|A_2|B_1|B_2|C_1|C_2|$ by interleaving the initial encoders parts without changing connections between adders and delay elements. The parameters of the combined encoder are $L^*=2L$, $k^*=2k$. Then we present the combined encoder in the form $|A^*|B^*|C^*|$ and define encoding stages similarly to (1).

The defect of a code sequence at the output of the combined encoder is not greater than 2*6 (for the second and third stages) as it follows from Lemma 1 and Theorem 3. Let us consider a generalization of the proposed code construction using the interleaving of encoders. The combined encoder of the f-th order has the parameters: $L^*=fL$, $k^*=fk=L^*/3$.

THEOREM 4. The combined encoder of the f-th order is an s-optimal encoder, where $s \geqslant ((f-1)/f)k^*-6f-1$.

Proof. All sequences I_l, $l \leqslant (f-1)k$, for this encoder have zero defect because the distance between the inputs of every adder is greater than the length of these sequences. The sequences I_l, $l>(f-1)k$ generate at least $2k^*+ min(2k^*,k^*+l-1)-6f$ nonzero output symbols, as it follows from Lemma 1 and Theorem 1 . ■

THEOREM 5. A combined encoder generates a non-catastrophic code.

Proof. Let us consider two adders of the combined encoder, connected with the two first delay elements of the encoder register. The outputs of these adders are equal to zeros if

$$\begin{cases} a_i + a_{i+k(f+1)} + a_{i+2kf+1} = 0 \\ a_{i+1} + a_{i+k(f+1)-1} + a_{i+2kf+2} = 0 , \end{cases} \quad (2)$$

where a_i is the i-th symbol of the information sequence. From (2) we have condition $a_j + a_{j+2} = 0$, which is valid for two information sequences only. One of them consists of the all-one sequence. Another sequence consists of interchanging zeros and ones. The all-one sequence does not generate the zero codeword because the adders have odd number of inputs. The second sequence does not satisfy the first equation of (2) because for this sequence $a_i + a_{i+2kf+1} = 1$ and the first equation becomes $a_{i+k(f+1)} = 1$. That is not true for each second symbol of this sequence. ■

Thus the proposed construction provides a parameter s that approaches its optimum value as the code alphabet grows in size.

References.

[1] A. J. Viterbi, I. M. Jacobs, "Advances in coding and modulation for noncoherent channels affected by fading, partial band and multiple access interference," Advances in communications systems, vol. 4, New York: Academic Press,pp. 279-308, 1975.
[2] E. M. Gabidulin "Convolutional codes over large alphabets," Proc. of the International Workshop on Algebraic and Combinatorial Coding Theory, Varna, Bulgaria, pp.80-84, 1988.
[3] K.A.S. Abdel-Ghaffar " Some convolutional codes whose free distances are maximal," IEEE Trans. Inform. Theory, vol.IT-35, pp.188-191, 1989.
[4] A. J. Viterbi, J. K. Omura "Principles of digital communications and coding," McGraw-Hill N.Y., 1979.

POSITION RECOVERY ON A CIRCLE BASED ON CODING THEORY

by
Benjamin Arazi
Dept. of Electrical and Computer Engineering
Ben Gurion University
Beer Sheva, ISRAEL

ABSTRACT

We treat the problem of designing periodical binary sequences for the purpose of being able to recover the position of a specified bit in the sequence in respect to a defined starting position. This can be useful for example for position recovery on a rotating device. Known sequences which can serve this purpose are the de-Bruijn sequences of periodicity 2^n or maximal sequences of periodicity $2^n -1$. Both such sequences have the property that any consecutive n bits in them form a different pattern, and therefore define uniquely their position with respect to a defined starting point. Although each bit in a maximal length sequence of periodicity $2^n -1$ starts a different n-tuple, the complexity of recovering the position of a given n-tuple, based on its specific pattern, is that of performing a log operation over $GF(2^n)$, which is exponential. Maximal length sequences are characterized by a structure under which location indices that belong to the same cyclotomic coset are assigned the same value. A new class of binary sequences of periodicity $2^n -1$, characterized by the same structure (location indices that belong to the same cyclotomic coset are assigned the same value) is presented in this paper. It is shown that in these sequences the recovery of a location index of any specified bit is of complexity that is *linear* with n. There is however an additional complexity defined as 'space complexity', which is related to the fact that the bits whose values determine the location of a specified bit are not all located in the direct vicinity of the bit whose location is to recovered. The resemblance between the space complexity encountered here, and that encountered in a standard binary search is discussed.

1. INTRODUCTION

1.1. The complexity of recovering the location index of a given n-tuple in a maximal length sequence of periodicity 2^n -1

Every bit in a maximal length sequence of periodicity 2^n -1 starts a different n-tuple. Relative to a specified starting point, the position of any given bit in such a sequence is therefore determined uniquely based on the value of this bit and other n-1 neighboring ones. The use of this property for position recovery on a rotating device has already been shown [1].

A log operation over GF(2^n) means the recovery of the exponent x, given a field element α^x. Such an operation is of exponential complexity [2]. Intuitively, the complexity of performing a log operation over a finite field raises from the impossibility of defining an order on a closed circle.

Theorem 1: Given an n-tuple *A* in a maximal length sequence of periodicity 2^n -1, determining its position with respect to an n-tuple *B* is equivalent in complexity to performing a log operation over the field.

Proof: Given α^x, where α is the root of a primitive polynomial g(t) of degree n, multiply an n-tuple *B* by $\alpha^x, \alpha^{x+1}, \ldots, \alpha^{x+n-1}$. The resultant bits form an n-tuple *A*, whose distance from *B* in the maximal length sequence generated by g(t), is x. Measuring distance between two n-tuples in a maximal length sequence is therefore equivalent in complexity to a log operation.

Theorem 2: Given $\alpha^x \in$ GF(2^n), the complexity of finding the parity of x is that of a log operation over GF(2^n).

Proof: A log operation over GF(2^n) can be performed by n operations in which the parity of x is found for different $\alpha^x \in$ GF(2^n). This is based on the simple fact that since operations in the exponent of α^x are performed modulo 2^n-1, the binary representations of the exponents of

$$\alpha^x, \ (\alpha^x)^2, \quad (\alpha^x)^{2^2}, \quad . \ . \ . \ . \ , (\alpha^x)^{2^{n-1}}$$

are cyclic shifts of the binary representation of x and their parity bits then form x.

1.2. Constructions based on cyclotomic cosets.

Definition: A cyclotomic coset modulo 2^n-1 of an integer $x \in [0, 2^n-2]$ consists of all the distinct numbers from the set $\{x, 2x \bmod(2^n-1), 4x \bmod(2^n-1),..., 2^{n-1}x \bmod(2^n-1)\}$.

Demonstration 1: The cyclotomic cosets modulo 31 are: {0}, {1,2,4,8,16}, {3,6,12,24,17}, {5,10,20,9,18}, {7,14,28,25,19}, {11,22,13,26,21}, {15,30,29,27,23}.

Definition: A sequence *B* of length m is obtained from a sequence *A* of the same length by h-decimation of *A*, if *B* is constructed by taking every h-th element of *A*, starting with the same element with which *A* starts, and repeating the process cyclically (the last element is considered to be followed by the first one) until m elements are obtained.

It should be noted that if (h, m) = 1, the elements of *B* consist of all the elements of *A*. If (h, m) > 1 then *B* consists of repetitions of a subset of *A*. This means, as a special case, that if the length of *A* is odd and h=2, then *B* consists of all the elements of *A*.

Notation: $\{C\}_n$ denotes the set of numerical sequences of periodicity 2^n-1, where the same values are assigned to locations whose indices belong to the same cyclotomic coset modulo 2^n-1.

Maximal length sequences are well known members of $\{C\}_n$ [3].

Demonstration 2: The general structure of a sequence $A \in \{C\}_5$ (based on the cyclotomic cosets listed in demonstration 1) is:

0	1	2	3	4	5	6	7	8	9	10	11	12	13	14	15	16	17	18	19	20	21	22	23	24	25	26	27	28	29	30
A	B	B	C	B	D	C	E	B	D	D	E	C	E	E	F	B	C	D	E	D	E	E	F	C	E	E	F	E	F	F.

The proof of the following theorem is self explanatory and is omitted.

Theorem 3: The sequence obtained by 2^i-decimation of $A \in \{C\}_n$ starting with location x, equals the sequence obtained by scanning *A* continuously (1-decimation) starting with location $(x/2^i) \bmod(2^n-1)$, for any $0 \leq x \leq 2^n-2$ and $0 \leq i \leq n-1$.

Demonstration 3: 4-decimating the sequence A of Demonstration 2, starting with location x=13, for example, yields the sequence ECEEFBC.... This same sequence is obtained by scanning A continuously, starting with location (13/4)mod 31 = 11.

Starting with x = 0, the 2^i-decimated sequence is identical to the continuous sequence starting at the same point. *If $A \in \{C_n\}$ is arranged in a circular form, its starting point is defined as the one starting with which all the sequences obtained by 2^i - decimation $i = 0, 1, ..., n-1$ are identical.* Note that a decimation process is naturally performed on a sequence arranged in a circular form since the process is cyclic by definition. We treat in this paper only those sequences with a single starting point as defined here.

1.3. A variation of a standard binary search, and introducing the concept of 'space complexity'.

Consider the problem of assigning different numerical values to N elements in an ordered array, for the purpose of recovering the location of any specified element, with respect to a defined starting point. The problem is trivially solved by assigning $\log_2 N$ bits to each element, representing its index in the range [0, N-1]. A binary search approach enables recovering the location index of an element by assigning to each element $\log_2(\log_2 N)$ bits. i.e., one value out of $\log_2 N$ possibilities.

In order to clarify the above statement, consider for example the following version (out of many) of a standard binary search.. Assign one unique value to all the elements having an odd location index. Another unique value will be assigned to all the elements having an index congruent to 2, when taken modulo 4, and in general: all the elements with location index x, for $x \bmod 2^i \equiv 2^{i-1}$, are assigned a unique value, $i=1,2,...,[\log_2 N]+1$. An additional unique value should be assigned to the first element (location index 0).

Demonstration 4: Take the case where N=13. The elements themselves are named A, B, C, D, E, F, G, H, I, J, K, L, M. Assigning respectively the values a, b, c, d for i=1, 2, 3, 4, and assigning e to location 0 yields the following:

Location of the element:	0	1	2	3	4	5	6	7	8	9	10	11	12
Name of the element:	A	B	C	D	E	F	G	H	I	J	K	L	M
Value assigned to the location:	e	a	b	a	c	a	b	a	d	a	b	a	c

The following process for recovering the location index of an element is based on an approach where pairs of values from the set {a, b, c, d, e} are checked for equality. No table look up (specifying that a is associated with odd indices etc.) is needed. We demonstrate how the location index x of the element L is recovered.

Step a: Recovering the parity of x; check whether the values assigned to location indices x and x-2 equal. If they do, x is 1. Otherwise it is 0. In our case the values assigned to locations x and x-2 are both a, and x is then 1.

If x is 1 then substitute x = x-1.

Go to step b.

Step b: Recovering the next significant bit of x; check whether the values assigned to location indices x and x-4 equal. If they do, the next significant bit of x is 1. Otherwise it is 0. In our case the values assigned to locations x and x-4 (x was reduced by 1 at the end of Step a and is now the location index of K) are both b, and the next significant bit of x is 1.

If the next significant bit of x is 1 then substitute x = x-2.

Go to step c.

Step c: Recovering the next significant bit of x; check whether the values assigned to location indices x and x-8 equal. If they do, the next significant bit of x is 1. Otherwise it is 0. In our case the values assigned to locations x and x-8 (x was reduced by 2 at the end of Step b and is now the location index of I) are d and e, and the next significant bit of x 0.

The process terminates when we get to e, which is the value assigned to location #0. (This is the only value stored for reference purposes.) If we arrive to e by subtracting 8 from x, then the most significant bit in x is 1. In out case we then have that $x = (1011)_2$.

Generally there are n-1 steps in the described process (for an array of length p, $2^{n-1} \leq p < 2^n$). If one arrives to location #0 before n-1 steps end, then the rest of the unrecovered more significant bits are all 0.

The problem of recovering the location index of an element in an ordered array of length N is of course trivially solved by assigning $\log_2 N$ bits to each element, representing its index in the range [0, N-1]. This approach is based on an Alphabet of size N. A binary search approach enables recovering the location index of an element by assigning to each element $\log_2(\log_2 N)$ bits. i.e., Alphabet of $\log_2 N$ elements. The above process is based on comparing values associated with locations that are 2^i places apart, $i = 1, 2, \ldots, n-1$. Other approaches to binary search will still need 'travelling' from location index x to index $x - 2^i$, even if the recovery of the bits of x is not based on comparing values. (Another possible approach which will still need the 'travelling' would be to refer to a list which associates the values a, b, c, ... with the parameters based upon which these values were initially assigned.) This 'travelling' introduces 'space complexity' which is traded for the fact that we use Alphabet of size $\log_2 N$ rather than size N.

One can argue that if we already pursue the described travelling procedure we might as well directly count the number of places from an unknown location to the starting point, thereby recovering the location index by brute-force counting. However, the space complexity considerations treated before relate to skipping n *fixed* distances rather than counting continuously up to 2^n single steps.

A binary search is sequential by definition. That is, the more significant bits in the binary representation of i are recovered only after recovering the least significant ones. This is also part of the trade-off discussed above.

1.4. Statement of the problems treated in this paper and brief description of the main results.

Problem #1: Using a maximal length sequence for position recovery on a circle, based on assigning one bit to each location, was shown in Section 1.1 to be of exponential complexity (the complexity of performing a log operation over $GF(2^n)$). Are there other binary sequences of periodicity $2^n - 1$ which enable position recovery on a circle based on lower complexity? Answer to be given in the paper: Yes. There are binary sequences which enable position recovery based on *linear* complexity, but which have space complexity of the form discussed in Section 1.3. Four values which are 2^i places apart in the sequence are to be compared for each recoverd bit in the binary representation of a location index.

Problem #2: Considering the fact that the structure of any sequence $A \in \{C\}_n$ (Section 1.2) is based on modular arithmetic operations, does this necessarily mean that position recovery in them is complex? (One may think that the answer is probably yes, as defining an order, or even a starting point, on a circle is unnatural.) Answer to be given in the paper: No. There are sequences $A \in \{C\}_n$ in which position recovery is based on *linear* complexity in time, but which have space complexity. Four values which are 2^i places apart in the sequence are to be compared, for each recoverd bit in the binary representation of a location index.

Problem #3: How complex is it to find the **parity** of the location index of an element in $A \in \{C\}_n$? (As shown in Section 1.1, this problem is of exponential complexity for maximal length sequences.) Answer to be given in the paper: There are sequences $A \in \{C\}_n$ in which the parity of location index x is recovered by performing trivial comparison operations on the values assigned to locations x, x+1, x+2, x+3.

Problem #4: Can a binary search of the form described in Section 1.3 be based on Alphabet of size smaller than $\log_2 N$? Answer to be given in the paper: Yes. Even Alphabet of *two* characters is sufficient, provided that four values which are 2^i places apart are compared, rather than two.

Problem #5: Can a binary search be executed in parallel? That is, can all the bits in the binary representation of the location index x in an ordered sequence be recovered simultaneously, and not serially which is an inherent property of the process treated in Section 1.3. Answer to be given in the paper: Yes. Values can be assigned to the elements in an ordered array (one bit per element is sufficient), based upon which all the bits in the binary representation of a location index x are recovered *simultaneously*, provided that four values which are 2^i places apart are compared, rather than two, for each recovered bit.

2. A BINARY SEQUENCE $B_n \in \{C\}_n$

2.1. Recovering the parity bit of the location index of an element in a binary $B_n \in \{C\}_n$

Notation: PHW(x) denotes the parity of the Hamming weight of the binary representation of x.

Notation: B_n denotes a binary sequence of $2^n - 1$ elements, whose i-th element, $i = 0, 1, \ldots, 2^n - 2$, is PHW(i).

Demonstration 5: The sequence B_5 is:

0 1 1 0 1 0 0 1 1 0 0 1 0 1 1 0 1 0 0 1 0 1 1 0 0 1 1 0 1 0 0.

Note that B_n is actually the last column of a Sylvester-ordered Hadamard matrix of order $2^n \times 2^n$, with its last element dropped.

Theorem 4: $B_n \in \{C\}_n$

Proof: If the binary representation of x consists of n bits then the binary representation of $y = (2^i x) \bmod 2^n - 1$ consists of that of x shifted cyclically to the left for i places. This means that if a and b are in the same cyclotomic coset modulo $2^n - 1$, then PHW(a) = PHW(b). It then follows that B_n has equal values in locations belonging to the same cyclotomic coset modulo $2^n - 1$.

Theorem 5: If PHW(x) = PHW(x+1) for a certain x, then x is odd.

Proof: For an even x it is clear that PHW(x) $\neq$ PHW(x+1), since the least significant bits in the binary representation of x and x+1 are 0 and 1 respectively, where the rest of the bits are the same in both representations. It follows that if PHW(x) = PHW(x+1) for a certain x, then x must be odd.

Theorem 6: Given at most four successive values in B_n, starting with index x, the parity of x is recovered by a deterministic process based on value comparisons.

Proof: Based on the fact that PHW(x) $\neq$ PHW(x+1) = PHW(x+2) $\neq$ PHW(x+3) for x mod 4 = 0, it follows that it is impossible to have more than two successive identical values in B_n. Having two successive identical values, the first out of the two has an odd location index based on Theorem 5. If the given group of four does not contain two successive equal values, that is, it is of the form a, a', a, a', then the following bit (in B_n) must be a', in which case the first bit in the given group of four has an even location index.

Conclusion 1: Let A, B, C, D, be four successive bits taken from B_n. The location index (in B_n) of A is odd if and only if A=B or C=D.

Fig. 1(a) depicts a circuit which generates the parity bit of the location index of A, given A, B, C, D.

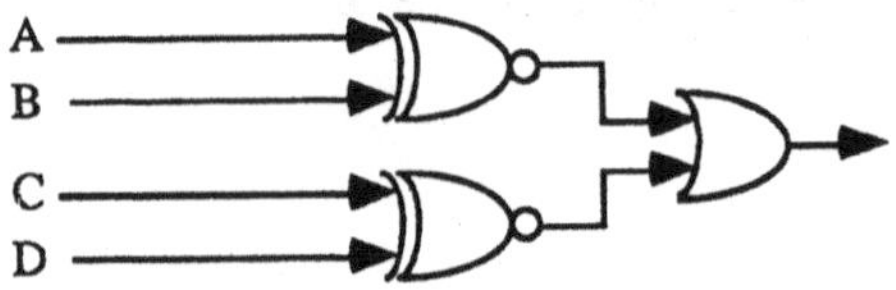

(a)

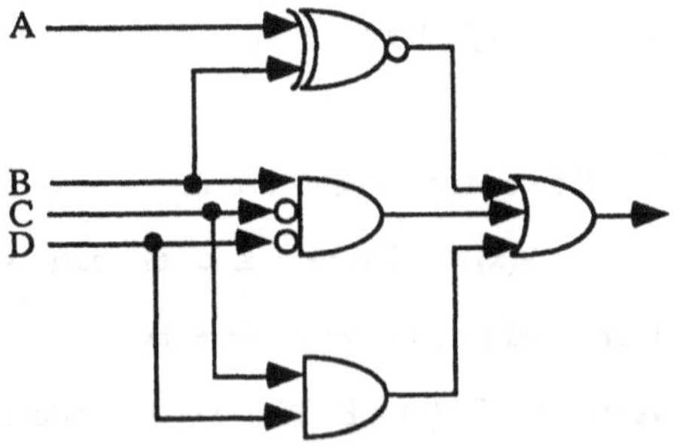

(b)

Fig 1. Recovering the parity bit of the location index of the element A in a cyclically arranged sequence B_n, given the four successive elements A, B, C, D. (a) Even n. (b) Odd n.

Consider the case where B_n is scanned cyclically. Conclusion 1 and the circuit of Fig. 1(a) are still valid for an even n. A case of three successive 0's will however occur for an odd n, starting with the place before last and continuing with the beginning of the sequence. (See the sequence B_5 listed above.) Here, the two equal successive bits which are consistent with Theorem 5 are the first two 0's. Fig. 1(b) depicts the recovery of the parity bit of the location index of the element A, for an odd n.

It should be noted that recovering the parity of the location index of the first bit out of a given group of n successive bits taken from a maximal length sequence of periodicity 2^n-1 can also be done by a logic circuit. However, the structure of such a circuit is not based on a deterministic process, whereas the circuits of Fig. 1 are based on a defined approach of bits comparisons.

2.2. Recovering the position of an element in a circularly arranged B_n

Observe the following features of B_n

(a) The elements of B_n in places x, (x+2) mod (2^n-1), (x+4) mod (2^n-1) and (x+6) mod(2^n-1) equal, respectively, the four successive elements in the sequence starting with place (x/2) mod (2^n-1). Generally: the elements of B_n in places x, $(x+2^i)$ mod (2^n-1), $(x+2^{i+1})$ mod (2^n-1) and $(x+3\cdot 2^i)$ mod (2^n-1) equal the four successive elements starting at place $(x/2^i)$ mod (2^n-1).

(b) The binary representation of $y = (x/2^i)$ mod (2^n-1) is obtained from that of x by shifting the latter cyclically for i places to the right. In other words: the parity bit of y is the i-th coefficient in the binary representation of x. In view of Theorem 6 and observation (a) above, the first four elements in the 2^i-decimated sequence B_n, starting with location x yield the coefficient of 2^i in the binary representation of x, i = 0, 1,...,n-1. This enables the recovery of x by n different decimations of B_n, starting with element #x.

Let B_n be arranged in a circular form. The starting point is defined as the one starting with which any 2^i - decimation yields the same sequence. (There is only one such a point.) The first four elements in the 2^i - decimated sequence B_n, starting with location x, yield the coefficient of 2^i in the binary representation of x, i = 0, 1,..., n-1. This enables the recovery of x by n different decimations of B_n, starting with element #x.

Demonstration 6: Fig. 2 depicts the sequence B_4 arranged in a circle (clockwise). Nine specific elements of the sequence are numbered in Fig 2(a). Starting with the location to be recovered, the first four elements in the 2^i-decimated sequence B_n, for i = 0, 1, 2, 3, are respectively: {①,②,③,④}, {①,③,⑤,⑥}, {①,⑤,⑦,⑨} and {①,⑦,②,⑧}. Let (x_3, x_2, x_1, x_0) denote the binary representation of the index to be recovered. These bits are recovered simultaneously by entering each group of four bits into the circuit of Fig. 1(a).

Recovering x_0. The values of elements ①,②,③,④ are 0,0,1,0, respectively. The output of the circuit of Fig. 1(a) is then 1. i.e., $x_0 = 1$.

Recovering x_1. The values of elements ①,③,⑤,⑥ are 0,1,1,0, respectively. The output of the circuit of Fig. 1(a) is then 0. i.e., $x_1 = 0$.

Recovering x_2. The values of elements ①,⑤,⑦,⑨ are 0,1,1,0. It follows that $x_2 = 0$.

Recovering x_3. The values of elements ①,⑦,②,⑧ are 0,1,0,0. The output of the circuit of Fig. 1(a) is then 1. i.e., $x_3 = 1$.

The binary representation of the recovered index is then 1 0 0 1, i.e., x = 9.

Claim: The total number of different elements of B_n, whose values are needed for recovering a position of an element, is 2n+1.

(As shown above, these 2n+1 values are partitioned into n nondisjoint sets of size 4, which is *fixed* for any n, where each set enables the recovery of one bit in the binary representation of the location index, simultaneously with the other bits.)

Proof: The first four elements in the 2^i - decimated sequence B_n, starting with location x, yield the coefficient of 2^i in the binary representation of x, i = 0, 1,..., n-1. This is summarized in the following list.

i (index of the recovered coefficient in the binary representation of x)	location of the four elements in B_n based upon which i is recovered. (All the additions are performed modulo $2^n - 1$.)
0	x, x+1, x+2, x+3
1	x, x+2, x+4, x+6
.	
.	
n-1	x, $x+2^{n-1}$, $x+2^n$, $x+3\cdot 2^{n-1}$

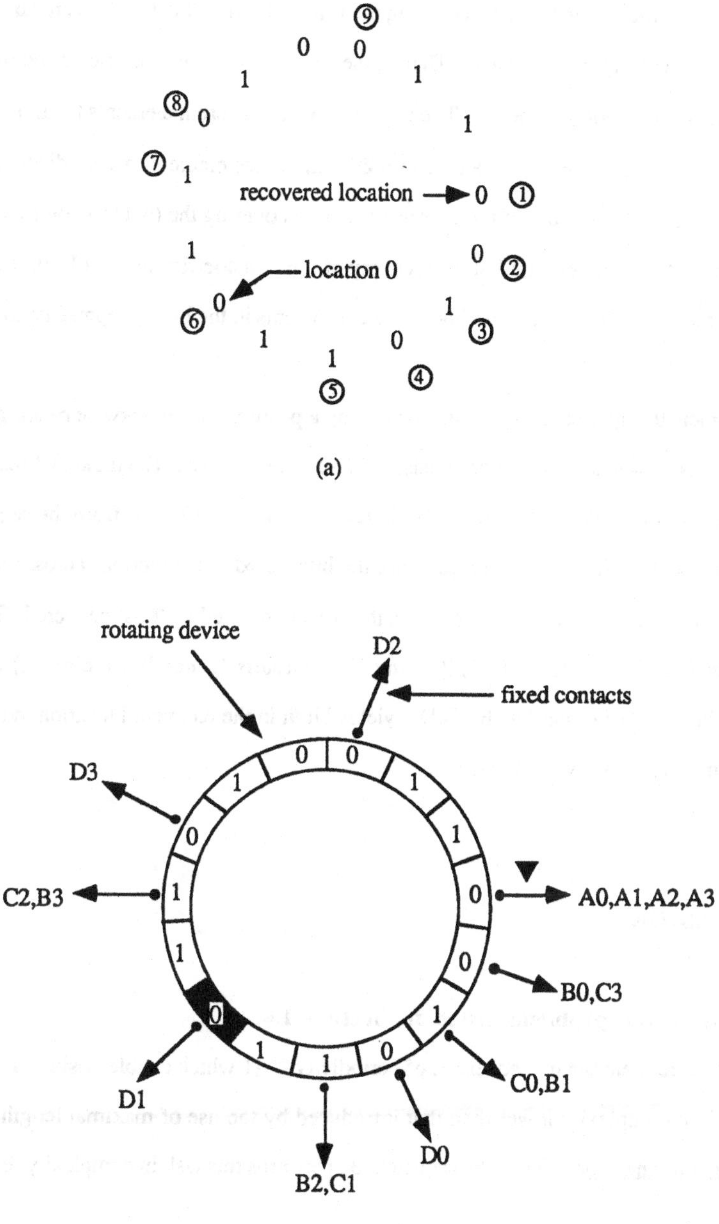

Fig. 2. Recovering a position in a circularly arranged B_4.

Out of the four elements needed for recovering the i-th coefficient, the first two are already used in the recovery of the (i-1)-th coefficient. That is, the recovery of a new coefficient requires at most two elements from B_n not used before. The total number of different elements from B_n needed for recovering x is then bounded by $4 + 2\cdot(n-1) = 2n + 2$. These elements are all distinct except for one case. The third element out of the four needed for recovering the (n-1)-th coefficient is $x+ 2^n$, which is the (x+1)-th element used in the recovery of the 0-th coefficient. 2n+1 different elements from B_n are then sufficient for recovering all the coefficients in the binary representation of x.

Fig. 2(b) depicts the application of B_n for recovering a position on a clockwise rotating device. In this specific case n = 4 and the device consists of $2^n-1 = 15$ sections. The darkened sector is sector #0, with respect to which the sections are indexed clockwise. One bit from the sequence B_4 is stored in each sector. Nine fixed contacts sense the bits stored in the sections across them. At any instant, the location index of the sector across the contact marked by ▼ is recovered. This is done by inputting {Ai,Bi,Ci,Di}, i = 0, 1, 2, 3 into four circuits (generally: n circuits) of the form depicted in Fig. 1. The group {Ai,Bi,Ci,Di} yields bit #i in the recovered location index (bit #0 is the LSB), simultaneously with the others.

3. DISCUSSION

3.1. Revising the problems listed in Section 1.4.

Problem #1: Are there binary sequences of periodicity $2^n -1$ which enable position recovery on a circle based on complexity lower than that introduced by the use of maximal length sequences? Answer given in the paper: Yes. The sequence B_n performs this task in complexity linear with n.

Problem #2: Is position recovery in any sequence $A \in \{C\}_n$, based on specified values in a sequence, necessarily complex? Answer given in the paper: No. Although $B_n \in \{C\}_n$, position recovery is still of linear complexity in time.

Problem #3: How complex is it to find the *parity* of the location index of an element in $A \in \{C\}_n$? Answer given in the paper: For B_n the parity of location index x is recovered by making trivial comparisons on the values assigned to locations x, x+1, x+2, x+3.

Problem #4: Can a binary search of the form described in Section 1.3 be based on Alphabet of size smaller than $\log_2 N$? Answer given in the paper: Yes. Using B_n, even Alphabet of *two* characters is sufficient, provided that four values which are 2^i places apart are compared, rather than two.

Problem #5: Can a binary search be executed in parallel? Answer given in the paper: Yes. By arranging B_n on a circle, all the bits in the binary representation of a location index x are recovered simultaneously, provided that four values which are 2^i places apart are compared, rather than two.

3.2. Observations related to a log operation over GF(2^n).

As pointed in Section 1.1 the complexity of finding the parity of the location index of an element of a maximal length sequence of periodicity 2^n-1 is that of a log operation over GF(2^n). On the other hand, it is trivial to find the parity of the location index of an element in B_n based on its value and the values of three neighboring bits. For a prime 2^n-1 there are (2^n-2)/n different maximal-length sequences, and the same number of cyclotomic cosets (excluding the coset {0}). It can then simply be shown that disregarding their first element, all the binary sequences of length 2^n-1 having equal values in locations belonging to the same cyclotomic coset modulo 2^n-1, including B_n, can be expressed as the sum (mod 2) of some maximal-length sequences. This observation illuminates the problem of performing a log operation over GF(2^n) from a new angle. Generally, whereas the concept 'the parity of the exponent of α^x' is highly artificial because parity is not defined when operating modulo 2^n-1 (every integer is divisible by 2 when operating modulo an odd integer), PHW(x) is consistent with both standard binary operations and modular arithmetic operations.

REFERENCES

[1] J.A. Bondy and U.S.R. Munty, "*Graph Theory with Applications*", American Elsevier, 1976.

[2] D. Coppersmith, "Fast Evaluation of Logarithms in Fields of Characteristic Two", IEEE Trans. Informat. Theory, vol. IT-30, pp. 587-594, 1984.

[3] S.W. Golomb, "*Shift Register Sequences,*" Holden Day, 1967.

THE LOWER BOUND FOR CARDINALITY OF CODES CORRECTING ERRORS AND DEFECTS

V.BLINOVSKY

WITH THE INSTITUTE FOR PROBLEMS OF INFORMATION TRANSMISSION
USSR ACADEMY OF SCIENCE

§ 1. Preface

Let $\mathbb{F}_2^n$ be the Hamming space of binary n-tuples, $d(\ ,\ )$ be the Hamming metric ; let $C_n(y,r) = \sum_{x\in\mathbb{F}_2^n:d(x,y)=r} x$ and $B_n(y,r) = \sum_{x\in\mathbb{F}_2^n:d(x,y)\leqslant r} C_n(y,r)$ be respectively the sphere and ball of radius r and center y.

Denote $W_n \doteq \{1,\dots,n\}$. For an arbitrary vector $x=(x_1,\dots,x_n)\in\mathbb{F}_2^n$ and $\mathcal{J}\subset W_n$ define $x_{|\mathcal{J}}=(x_{i_1},\dots,x_{i_{|\mathcal{J}|}})$, where $i_1<i_2<\dots<i_{|\mathcal{J}|}$, $i_j\in\mathcal{J}, j=\overline{1,|\mathcal{J}|}$. Define the function $\varphi(\ ,x_{|\mathcal{J}}):\mathbb{F}_2^n\to\mathbb{F}_2^n$

$$\varphi(y,x_{|\mathcal{J}})=(z_1,\dots,z_n);z_i=\begin{cases} y_i & ,i\in W_n\backslash\mathcal{J} \\ x_i\ , & i\in\mathcal{J}\end{cases}$$

The channel with m defects is the set $\Phi_m=\{\varphi(\ ,x_{|\mathcal{J}});\mathcal{J}\subset W_n\ ,\ x_{|\mathcal{J}}\in\mathbb{F}_2^{|\mathcal{J}|}, |\mathcal{J}|=m\}$. In other words whenever one uses the channel with m defects for transmitting an n-tuple $y\in\mathbb{F}_2^n$, this channel proves to be in some 'state' $\varphi(\ ,x_{|\mathcal{J}})\in\Phi_m$ and the received n-tuple is $\varphi(y,x_{|\mathcal{J}})$ .The channel with defects at first was investigated in [1].

Define a code (linear code) $A(n,k)$ of cardinality $|A(n,k)|=2^k$ as a subset (subspace) $A(n,k)\subset\mathbb{F}_2^n$; $R=k/n$ is the rate of the code $A(n,k)$.

We shall use the following definitions . Code $A(n,k,m)$ allows the transmission of N messages and corrects $m<n$ defects if there exist a coding function $\varrho(i,\varphi) :\overline{1,N}\times\Phi_m\to A(n,k,m)$ and a decoding function $\lambda(x):\mathbb{F}_2^n\to\overline{1,N}$, which satisfy the relation

$$\lambda(\varphi(\varrho(i,\varphi),x_{|\mathcal{J}})) = i\ ,\varphi\in\Phi_m\ ,x_{|\mathcal{J}}\in\mathbb{F}_2^n\ .|\mathcal{J}|=m,\ i=\overline{1,N}\ . \quad (1)$$

When such functions exist and the following equality is true

$$\lambda(\varphi(\varrho(i,\varphi),x_{|\mathcal{J}})+z) = i \;;\varphi\in\Phi_m ,x_{|\mathcal{J}}\in\mathbb{F}_2^n .|\mathcal{J}|=m,\ i=\overline{1,N} ,\quad (2)$$
$$z\in\mathbb{F}_2^n ,d(z,0)=t,$$

we say that the code $A(n,k,m,t)\subset\mathbb{F}_2^n$ can be used for transmitting N messages and correcting m defects and t errors.

Suppose that we need to transmit the message number i. The encoder 'knows' the defect $\varphi(\ ,x_{|\mathcal{J}})$ and encodes the message i to $\varrho(i,\varphi)\in A(n,k)$. Then the outer sequence $\varphi(\varrho(i,\varphi),x_{|\mathcal{J}})$ of the channel Φ_m is to be sent through the ordinary channel, where t errors can occur. The final outer sequence is the n - tuple $\tilde{z}=\varphi(\varrho(i,\varphi),x_{|\mathcal{J}})+z \;;z\in\mathbb{F}_2^n, |z|=t$. It is easy to see that λ is the decoding function which outputs the sequence $\tilde{z}$ to message i.

In this paper a lower bound for N is under consideration. In proving the statements methods of random coding are used. Denote the complexity of the construction of the code $A\subset\mathbb{F}_2^n$ as the volume of the ensemble from which the code A and the functions φ,ϱ are chosen.

It was shown in [2] that there exist codes which can be used to transmit N messages, correct m defects and t errors with parameters N,m,t satisfying the inequality

$$\log N/n>(1-m/n)(1-H(2t/(n-m)) -o(1). \quad (3)$$

The construction complexity of such codes is $\sim 2^{2^{O(n)}}$. In this work we show that under constraint (4) the bound (3) is true for codes with construction complexity $\sim 2^{O(n^{2+\varepsilon})}$, $\varepsilon\in(0,1/2)$.

We need the following notions. For an arbitrary set of vectors $\{g\}\subset\mathbb{F}_2^n$ denote by $\Lambda(\{g\})\subset\mathbb{F}_2^n$ the linear hull of $\{g\}$, let $D^L(\{g\})=\{x\in\mathbb{F}_2^n: d(x,\{g\})<L\}$.

Denote by $\{s_p(n)\}, p=1,.,5$ the sequences of positive integers, $s_p+s=o(n), \min(\ln(s_p),\ln(s))/\ln(n)\to\sigma>1/2, n\to\infty$, s_4 is even. Later we state the relations between s_p with different indexes p; s without index present in general formulas, its value is specified in every concrete case. All following formulas are true for sufficiently large values of n. Fix $\varepsilon\in(0,1/2)$.

The main result of this paper is formulated in the following

Theorem 1. Let $m,n: 0<m,t<n$ $t/n \to \tau>0$, $m/n \to \mu \in [0,1)$, satisfy the unequality

$$4t/n<(1-m/n)^2-s_1/n \quad , \quad (4)$$

then there exist codes, which asymptoticaly achieve the bound (3) and have construction complexity less than $2^{O(n^{2+\varepsilon})}$, $\varepsilon \in (0,1/2)$.

The asymptotical reduction of the construction complexity as compared with [2] is due to the technique of choosing the set of the linear codes, which generates the covering of the sets of special configuration in the Hamming space $\mathbb{F}_2^n$.

§ 2 . Proof of theorem 1.

Define the sequence of sets $\{Q(n)\} \subset \mathbb{F}_2^n$, $|Q(n)|=2^{nq(n)}$. For arbitrary $x \in \mathbb{F}_2^n$ define $Q(n,x)=Q(n)+x$. The following lemmas are valid .

Lemma 1. For a proportion of the linear codes $A(n,k) \subset \mathbb{F}_2^n$ not less than $1-2^{-s^2/2}$ and for arbitrary $x \in \mathbb{F}_2^n$ the following estimation is true

$$|Q(n,x) \cap A(n,k)| < (|Q(n)|\,|A(n,k)|\,2^{-n}+1)2^s \quad . \quad (5)$$

Assign ensemble $\mathcal{A}(n,k)$ of binary linear codes $A(n,k) \subset \mathbb{F}_2^n$ generator matrix $G(n,k)$ of dimension $n \times k$; the binary elements for the matrix are chosen independently and with equal probability. Denote by $E(\ )$ the mathematical expectation . Let's label vectors of every code from $\mathcal{A}(n,k)$ with numbers $1,..,2^k$ arbitrary , but in the same manner for all codes; fixed zero vector is $a_1 \in A(n,k)$.

For all $a_j \in A(n,k)$ let $\Gamma_j(x)=\{A(n,k) \in \mathcal{A}(n,k): a_j \in Q(n,x)\}$, $j=\overline{1,|A(n,k)|}$; λ_j -indicator of the set $\Gamma_j(x)$; $\lambda = \sum_{j=1}^{|A(n,k)|} \lambda_j$.

Assume, that the following relation is true for some positive integer $f=o(n)$

$$2^{k-f}E(\lambda_2)>1 \quad .$$

The following relation is a consequence of the definition of $\mathcal{A}(n,k)$

$$E\Big(\sum_{j=1}^{2^k} \lambda_j\Big)^f < (2^k E(\lambda_2))^f + (2^k E(\lambda_2))^{f-1} 2^{f-1} C_f^1 + \ldots + \tag{6}$$

$$+(2^k E(\lambda_2))^{f-i} 2^{i(f-i)} C_f^i + \ldots + 2^{k+f-1} E(\lambda_2) <$$

$$< (2^k E(\lambda_2))^f (1+2^{f-k}/E(\lambda_2))^f .$$

Using the estimation (6) and Chebyshev's unequality , one obtains

$$P(\lambda > 2^s E(\lambda)) = P(\lambda^f > 2^{sf} E^f(\lambda)) < P(\lambda^f > 2^{sf} E(\lambda^f) \times \tag{7}$$

$$\times (1+2^{f-k}/E(\lambda_2))^{-f}) < (1+2^{f-k}/E(\lambda_2))^f 2^{-sf} .$$

Let f=s .In the case $2^{k-f}E(\lambda_j)>1$ statement of lemma 1 follows from (7) and proper choice of f . Notice , that $E(\lambda_2)=|Q(n)|2^{-n}$ .In the case $2^{k-f}E(\lambda_2)<1$, choose $Q'(n)\subset\mathbb{F}_2^n$, $Q(n)\subset Q'(n)$, such, that $1<2^{k-f}|Q'(n)|<2^s$.Now we can prove lemma 1 for the new set $Q'(n)$, instead of $Q(n)$ and obtain that $|Q'(n,x)\cap A(n,k)|<(|Q(n)||A(n,k)|2^{-n}<2^s$ with probability not less, that $1-2^{-s^2/2}$. But $Q(n)\subset Q'(n)$, and $Q(n,x)\cap A(n,k)\subset Q'(n,x)\cap A(n,k)$, $|Q(n,x)\cap A(n,k)|<|Q'(n,x)\cap A(n,k)|$. This completes the proof of lemma 1.

We use next covering lemma (cf.[3]) .

Lemma 2.Let $|Q(n)||A(n,k)|2^{-n-2s}>1$.For a proportion of codes $A(n,k)\in\mathcal{A}(n,k)$ not less that $1-2^{-s^2/2}$, and for x uniformly chosen in $\mathbb{F}_2^n$ the following unequality is true

$$|D^{[(sn)^{1/2}]}(Q(n,x))\cap A(n,k)|>|Q(n)||A(n,k)|2^{-n-s} . \tag{8}$$

Let r=[2mt/(n-m)].Choose **max** k_1 and **min** k_2 which satisfy the following relations

$$2^{k_1+k_2-n+s_2} \qquad |C_n(0,2t+r)|<1 , \tag{9}$$

$$2^{k_2-m-s_3} \qquad |C_m(0,r)|>1 \quad . \qquad (10)$$

Define the ensemble $\mathcal{U}(n,k_1+k_2)=\mathcal{A}(n,k_1+k_2)\times\ldots\times\mathcal{A}(n,k_1+k_2)$ ($[n^{\varepsilon}]$ times , $\varepsilon\in(0,1/2)$) . Ensemble $\mathcal{U}(n,k_1+k_2)$ consists of $\sim 2^{O(n^{2+\varepsilon})}$ elements. We'll show later, that there exists element $U(n,k_1+k_2)\in \mathcal{U}(n,k_1+k_2)$, such that the union $u(n.k_1+k_2) = \bigcup_{i=1}^{[n^{\varepsilon}]} A^i(n,k_1+k_2)$ of the linear codes $A^i(n,k_1+k_2)$ - the components of $U(n,k_1+k_2)$, form a code which can be used to transmit 2^{k_1} messages and correct m defects and t errors .

Let $A(n,k_1)\subset A(n,k_1+k_2)$ ($A(n,k_2)\subset A(n,k_1+k_2)$) be the linear hull of the first k_1 (last k_2) rows of generating matrix of the code $A(n,k_1+k_2)$. Let's index vectors of $A(n,k_1)$, using numbers $1,..,2^{k_1}$; fixed zero vector is $a_1\in A(n,k_1)$.

Fix $U(n,k_1+k_2) = A^1(n,k_1+k_2)\times,\ldots,\times A^{[n^{\varepsilon}]}(n,k_1+k_2)$.

Associate to every message $i=\overline{1,2^{k_1}}$ with the set

$$\alpha_i=\{a_i^1,\ldots,a_i^{[n^{\varepsilon}]}\};\ a_i^l\in A^l(n,k_1), a_i^l\neq a_j^l, i\neq j.$$

For defect $\varphi(\ ,x_{|\mathcal{J}})\in\Phi_m$ consider the set

$$\alpha_i(\varphi)=\{a_i^1(\varphi),..,a_i^{[n^{\varepsilon}]}(\varphi)\}, a_i^l(\varphi)=\{a\in A^l(n,k_2)+a_i^l: |a_{|\mathcal{J}}+x_{|\mathcal{J}}| < $$
$$<r+s_4/2\ \}; i=\overline{1,2^{k_1}}\ ,l=\overline{1,[n^{\varepsilon}]}\ . \qquad (11)$$

For transmitting the message number i, when defect $\varphi(\ ,x_{|\mathcal{J}})$ occurs , we choose one vector from $D^{s_4/2}(\alpha_i(\varphi))$. Next we show that for arbitrary $i\in\overline{1,2^{k_1}}$, $\varphi(\ ,x_{|\mathcal{J}})\in\Phi_m$ $\alpha_i(\varphi)$ is nonempty and there exists vector $a_i(\varphi)\in D^{s_4/2}(\alpha_i(\varphi))$, such that the distance between $\varphi(a_i(\varphi),x_{|\mathcal{J}})$ and arbitrary vector from T(i)(see (12), set T(i) consist of the vectors , which are used for transmitting messages, differ from i) exceed $2t+r+2s_4$. At the same time from the definition of $\varphi(\ ,x_{|\mathcal{J}})$ and $\alpha_i(\varphi)$ follows that the distance between $\varphi(a_i(\varphi),x_{|\mathcal{J}})$ and $a_i(\varphi)$ does not exceed $r+s_4$. Consequently the ball $B_n(\varphi(a_i(\varphi),x_{|\mathcal{J}})+z,t+r+s_4)$, $z\in\mathbb{F}^n{}_2$, $|z|=t$, contains at least one codeword from $\alpha_i(\varphi)$, but nothing from T(i).This fact allows us to determine the decoding function as in (13).

Lemma 3. Under the conditions (4), (9), (10) there exists an element $U(n,k_1+k_2)\in\mathcal{U}(n,k_1+k_2)$, that for arbitrary message $i=\overline{1,2^{k_1}}$ and defect $\varphi(\ ,x_{|\mathcal{J}})\in\Phi_m$ there exists vector $a_i(\varphi)\in\alpha_i(\varphi)\subset u(n,k_1+k_2)$,which satisfies the following unequality

$$d(\varphi(a_i(\varphi),x_{|\mathcal{J}}),a) > 2t+r+2s_4 \quad , \qquad (12)$$

$$a\in T(i) = u(n,k_1+k_2)\setminus \bigcup_{j=1}^{[n^{\varepsilon}]} \{A^j(n,k_2)+a_i^j\} .$$

Let

$$\varrho(i,\varphi)=a(i,\varphi) \ ; \ \lambda(z) = \arg\min_{x\in u(n,k_1+k_2)} d(x,z) . \qquad (13)$$

It is easy to see,that the function $\lambda(z)$,defined in (13) realizes the decoding of the vector $z\in\mathbb{F}_2^n$ to the nearest vector from $u(n,k_1+k_2)$ Unequality (12) means that arbitrary vector $a\in u(n,k_1+k_2)$, exept the vectors from the set $\bigcup_{j=1}^{[n^{\varepsilon}]} \{A^j(n,k_2)+a_i^j\}$ which can be used for transmitting the message number i ,is farer than $2t+r+2s_4$ from $\varphi(a_i(\varphi),x_{|\mathcal{J}})\in\mathbb{F}_2^n$. The consequence of (12) is that when $|z|=t$,functions ϱ,λ , defined in (13) satisfy the relation (2) , and the code $u(n,k_1+k_2)$ can be used to transmit $|A(n,k_1)|=2^{k_1}$ messages and correct m defects and t errors.The estimation (3) can be easily obtained from (9) and (10) by choosing $r=[2tm/(n-m)n]$. Note, that $|\mathcal{U}(n,k_1+k_2)|=2^{O(n^{2+\varepsilon})}$. Hence we construct the code $u(n,k_1+k_2)$ as required in theorem 1.To complete the proof of theorem 1 we need to prove lemma 3.

Fix the message $i\in\overline{1,2^{k_1}}$ and defect $\varphi(\ ,x_{|\mathcal{J}})\in\Phi_m$.Denote $\beta(\mathcal{J},x_{|\mathcal{J}}) =\{\varphi(y,x_{|\mathcal{J}});y\in\mathbb{F}_2^n\}$,$\overline{\beta}(\mathcal{J},x_{|\mathcal{J}})=\{\beta(\mathcal{J},z_{|\mathcal{J}});z_{|\mathcal{J}}\in B_m(x_{|\mathcal{J}},r)\}$.It is easy to see,that $|\beta(\mathcal{J},x_{|\mathcal{J}})|=2^{n-m}$, $|\overline{\beta}(\mathcal{J},x_{|\mathcal{J}})|=2^{n-m}\sum_{j=0}^{r} C_m^j$.

We give an unformal proof of lemma 3; using the following comments one can reconstruct the whole proof .

Denote by $E(\ |\ )$ the conditional mathematical expectation , $\gamma=D^{s_4/2}(\bar{\beta}(\mathcal{J},x_{|\mathcal{J}}))$ and as before $a_i^l(\varphi)=(A^l(n,k_2)+a_i^l)\cap\gamma$.

At first we prove that with probability not less than $1-2^{-s_4 n^{\varepsilon}/4}$, for some $l\in\overline{1,[n^{\varepsilon}]}$, the intersection of the sets $a_i^l(\varphi)$ and $D^{2t+r+5s_4/2}(A^l(n,k_2))+a_j^l$ for all $j\neq i$ is empty (at the end of the paper we choose s_4, such that $s_4 n^{\varepsilon}>n^{1+\varepsilon_1}$ for some $\varepsilon_1\in(0,1/2)$). The last statement means that for some $l\in\overline{1,[n^{\varepsilon}]}$ (12) carried out for all $a_i(\varphi)\in a_i^l(\varphi)$ if $a\in A^l(n,k_1+k_2)\setminus(A^l(n,k_2)+a_i^l)$.

Then we prove, that for such l, there exists a nonempty subset $\tilde{a}\in D^{s_4/2}(a_i^l(\varphi))$ such that (12) is true for all $a_i(\varphi)\in\tilde{a}$ and $a\in u(n,k_1+k_2)\setminus A^l(n,k_1+k_2)$. Combining these two facts gives the proof of the lemma 3.

We have

$$\tilde{E}\doteq E(|\varphi(a_i^l(\varphi),x_{|\mathcal{J}})\cap\bigcup_{j\neq i}(D^{2t+r+5s_4/2}(A^l(n,k_2))+a_j^l)|)=$$

$$=\sum_{A\subset\mathbb{F}_2^n;\,b\in\mathbb{F}_2^n} E(|\varphi(a_i^l(\varphi),x_{|\mathcal{J}})\cap\bigcup_{j\neq i}(D^{2t+r+5s_4/2}(A^l(n,k_2))+a_j^l)|\,|A^l(n,k_2)=A,a_i^l=b)P(A^l(n,k_2)=A,a_i^l=a),$$ where $P(A^l(n,k_2)=A,a_i^l=b)=$

$=P(A^l(n,k_2)=A)2^{-n}$ and

$$E(|\varphi(a_i^l(\varphi),x_{|\mathcal{J}})\cap\bigcup_{j\neq i}(D^{2t+r+5s_4/2}(A^l(n,k_2))+a_j^l)|\,|A^l(n,k_2)=A,$$

$$a_i^l=b)\leq\sum_{j\neq i}E(|\varphi((A+a)\cap\gamma,x_{|\mathcal{J}})\cap(D^{2t+r+5s_4/2}(A)+a_j^l)|=$$

$$=|\varphi((A+a)\cap\gamma,x_{|\mathcal{J}})|\,|D^{2t+r+5s_4/2}(A)|2^{k_1-n}\leq 2^{O_1(s_4)}|(A+a)\cap\gamma|C_n^{2t+r}\times$$

$$\times 2^{k_1+k_2-n}.$$

Here we use the independence of pairs $a_i^l;a_j^l$, $j\neq i$; independence $A^l(n,k_2)$ from a_i^l and estimation

$C_n^{2t+r+5s_4/2} < C_n^{2t+r} 2^{-3s_4 \log_2 (t/n)}$. Accoding to lemma 1 and conditions (9),(10) for arbitrary $a \in \mathbb{F}_2^n$, $|(A+a) \cap \gamma| <$ $< |A| \, |\bar{\beta}(\mathcal{J}, x_{|\mathcal{J}})| 2^{0_2(s_4)-n} < 2^{0_2(s_4)+s_3/2}$ with probability not less, than $1-2^{-s_3^2/8}$.

Suppose, that $s_2 > 2(0_1(s_4)+0_2(s_4)+s_3/2)$; then

$$\tilde{E} < \sum_{\substack{A \in \mathbb{F}_2^n: \\ |(A+a)\cap\gamma| < 2^{s_3/2+0_2(s_4)}}} 2^{o_1(s_4)+k_1+k_2-n} \, C_n^{2t+r} |(A+a)\cap\gamma| P(A^1(n,k_2)=A) \; +$$

$$+ 2^{-s_3^2/8} \quad < 2^{-s_2/3} \; .$$

Here we use the estimation (9). Using Chebyshev's inequality for fixed $l \in \overline{1,[n^\varepsilon]}$ we obtain

$|\varphi(a_i^1(\varphi), x_{|\mathcal{J}}) \cap \bigcup_{j \neq i} (D^{2t+r+5s_4/2}(A^1(n,k_2)) + a_j^1)| = 0$ with probability

not less than $1-2^{-s_2/4}$.

Accoding to the definition of the ensemble $\mathcal{U}(n,k_1+k_2)$, different codes $A^1(n,k_1+k_2)$ are chosen independently. Consequently with probability not less than $1-2^{-s_2 n^\varepsilon/4}$ there exists $l \in \overline{1,[n^\varepsilon]}$ such that

$$|\varphi(a_i^1(\varphi), x_{|\mathcal{J}}) \cap \bigcup_{j \neq i} (D^{2t+r+5s_4/2} (A^1(n,k_2)) + a_j^1)| = 0 \; . \qquad (14)$$

Next we show, that for this l there exists a vector $a_i(\varphi) \in a_i^1(\varphi)$, which satisfies (12) with $a \in u(n,k_1+k_2) \setminus A^1(n,k_1+k_2)$.

Let $u^1(n,k_1+k_2) = u(n,k_1+k_2) \setminus A^1(n,k_1+k_2)$. We are going to estimate the number of vectors $x \in \beta(\mathcal{J}, x_{|\mathcal{J}})$, which satisfy the inequality $d(x,y) < 2t+r+5s_4/2$ for some $y \in u^1(n,k_1+k_2)$. The following relations are true

$$|D^{2t+r+5s_4/2} (u^1(n,k_1+k_2)) \cap \beta(\mathcal{J}, x_{|\mathcal{J}})| < \sum_{z \in u^1(n,k_1+k_2)} \qquad (15)$$

$$|B_n(z,2t+r+5s_4/2)\cap\beta(\mathcal{J},x_{|\mathcal{J}})| =$$

$$= \sum_{f=0}^{m} \sum_{\substack{z\in u^1(n,k_1+k_2):\\ d(z_{|\mathcal{J}},x_{|\mathcal{J}})=f}} |B_n(z,2t+r+5s_4/2)\cap\beta(\mathcal{J},x_{|\mathcal{J}})| < 2^{2O_3(s_4)} \times$$

$$\times \sum_{f=0}^{m} (1+2^{k_1+k_2-m}|C_m(0,f)|)C_{n-m}^{2t+r-f}+2^{-(O_3(s_4))^2/2} < 2^{3(O_3(s_4))} \cdot$$

$$(C_{n-m}^{2t+r}+2^{k_1+k_2-m}C_n^{2t+r}).$$

Second inequality in the last relations comes from lemma 1 and independence of the $A^j(n,k_1+k_2)$ with different j ; last inequality is a consequence of the condition (4) and equality $r=[2tm/(n-m)]$. Here we also use the estimations $H(1/2-\xi) < 1-4\xi^2/\ln 2, \zeta\in(-1/2,1/2)$ and $|B_{n-m}(0,2t+r+5s_4/2)-f)| < 2^{O_3(s_4)}C_{n-m}^{2t+r}$. Let $s_1 > (3nO_3(s_4))^{1/2}, s_2 > 6(O_3(s_4))$. From (4),(9) and (15) obtain

$$|D^{2t+r+5s_4/2}(u^1(n,k_1+k_2))\cap\beta(\mathcal{J},x_{|\mathcal{J}})| < 2^{n-m-s_2/3} \quad .(16)$$

Next relation follows from (16)

$$|y\in\bar\beta(\mathcal{J},x_{|\mathcal{J}}):\varphi(y,x_{|\mathcal{J}})\in D^{2t+r+5s_4/2}(u^1(n,k_1+k_2))\cap\beta(\mathcal{J},x_{|\mathcal{J}})| < \quad (17)$$
$$< 2^{n-m-s_2/3}|B_m(0,r)|.$$

Inequality (17) means, that with probability not less than $1-2^{-(O_3(s_4))^2/2}$, the number of vectors from $\bar\beta(\mathcal{J},x_{|\mathcal{J}})$, which can't be included in $a_i(\varphi)$ out of the distance between any of them and $u(n,k_1+k_2)$ is less, than $2t+r+5s_4/2$, is small compared to $|\bar\beta(\mathcal{J},x_{|\mathcal{J}})|$.

Consider the set

$$\delta^1=\bar\beta(\mathcal{J},x_{|\mathcal{J}})\setminus\{y:\varphi(y,x_{|\mathcal{J}})\in D^{2t+r+5s_4/2}(u^1(n,k_1+k_2))\cap\beta(\mathcal{J},x_{|\mathcal{J}})\}. \quad (18)$$

Now reduce $2t+r+5s_4/2$ to $2t+r+2s_4$ in all previous

formulas, where $2t+r+5s_4/2$ occur. This procedure does not change any of the previous estimations. Denote $\gamma^1 = D^{s_4/2}(\delta^1) \cap (A^1(n,k_2) + a_i^1) \subset a_i^1(\varphi)$. Let $s_4/2 > (ns_3)^{1/2}$. From lemma 2 and relation (17) follows

$$P(|\gamma^1| > 2^{s_3/3}) > \sum_{S \subset \mathbb{F}_2^n : |S| > 2^{n-m}(1-2^{-s_2/3}) C_n^r} P(\delta = S, |D^{s_4/2} S \cap (a_i^1 + A^1(n,k))| > 2^{s_3/3}) > (1-2^{-(O_3(s_4))^2/2})(1-2^{-s_3^2/2}) . \tag{19}$$

Here we use the independence δ^1 and $a_i^1 + A^1(n,k_2)$. Set γ^1 consists of vectors from $a_i^1(\varphi)$ such that $\varphi(\gamma^1, x_{|\mathcal{J}})$ is further than $2t+r+2s_4$ from $u(n,k_1+k_2) \setminus A^1(n,k_1+k_2)$. At the same time from (19) follows, that the set γ^1 is nonempty for all l. It follows from (14), that for some $l \in \overline{1,[n^\varepsilon]}$ the set γ^1 is further, than $2t+r+2s_4$ from $A^1(n,k_1+k_2) \setminus (A^1(n,k_2) + a_i^1)$. Combining this two facts and choosing $a_i(\varphi) \in \gamma^1$, we satisfy condition (12). This completes the proof of the lemma 3.

It is left to repeat the choice of s_p for different p. Choose s_p with properties as pointed out in preface.

1. Choose s_3.
2. Choose even $s_4 > 2(ns_3)^{1/2}$ and $s_4 > n^{1-\varepsilon+\varepsilon_1}$ for some $\varepsilon_1 > 0$.
3. For assigned in the text $O_1(s_4), O_2(s_4), O_3(s_4)$, choose $s_1 > (3nO_3(s_4))^{1/2}, s_2 > 6(O_3(s_4))$;

for that values s_1 and s_2 parameters k_1, k_2, m, t must satisfy the conditions of theorem 1.

Open problem. Prove theorem 1, omitting the restriction (4).

REFERENCES

1. B.S.Tsybakov,A.V.Kuznetsov.Coding in memory with defect cells .Probl.Inf.Trans.1974.v.10.№ 2.pp.52-60.
2. L.A.Bassalygo,M.S.Pinsker.Correcting of errors in channels with defects.Doklad.Akad.Sci.USSR.1987.v.243.№ 6.pp.1361-1364.
3. V.M.Blinovsky. Asymptoticaly tight uniform bounds for spectrum of cosets of the linear codes.Probl.Inf.Trans.1990. v.26.№1.pp.99-103.
4. V.Blinovsky. Covering of Hamming space by translations of the sets by vectors of linear code .Probl.Inf.Trans.1990. v.26.№3.pp.21-28.

SOFT DECODING FOR BLOCK CODES OBTAINED FROM CONVOLUTIONAL CODES

Boris D. Kudryashov
Leningrad Aircraft Equipment Institute
Hertsen str. 67, 190000 Leningrad USSR

Abstract. Bounds for the minimum distance and the error probability of decoding for tail biting convolutional codes are presented. We propose a decoding algorithm for these codes and obtain a bound for the asymptotical complexity of almost maximum likelihood decoding.

1. Introduction.

In this paper we analyze the performances of block codes obtained from convolutional codes. We consider generalized tail biting convolutional codes introduced by Ma and Wolf [1]. These codes are better than conventional terminated convolutional codes investigated before [2]. Two main problems concerning the construction of block codes are:

- to find codes and decoding algorithms with good relationships between the error probability and the decoding complexity;
- to propose a decoding algorithm of minimum complexity for asymptotically good block codes , i.e. for codes satisfying to Varshamov-Gilbert bound. The error probability of decoding has to be close to that for maximum likelihood decoding.

Both problems were considered in various works, but mainly for binary symmetric channel only. The first problem has been discussed by Kolesnik and Zigangirov [3]. The second one has been investigated by Evseev [4], Barg and Dumer [5] and Krouck [6]. Some results concerning the second problem for the case when soft decoding is admissible were obtained by Wolf [7] and Dumer [8].

Further advances in both directions were obtained by employment of tail biting convolutional codes. Some results are presented in this paper. Partially they were published in [9,10].

2. Tail biting codes.

The encoder for a block code obtained from a convolutional code by terminating is presented in Fig. 1. The initial convolutional encoder has constraint length ν =2, branch length (number of outputs) n_0= 2, rate r = $1/n_0$ = $1/2$. (For simplicity we consider only codes of rate

$1/n_0$). To produce one code word of block code we initialize encoding with all-zero state. Then k binary messages and ν zeros enter the encoder register. The output of the encoder is a sequence of $n = (k+\nu)n_0$ code symbols. The set of all these sequences is a linear (n,k)-code of rate $R = k/n = k/(k+\nu) * (1/n_0)$. For example, encoder presented in Fig.1 provides $(12,4)$-code with minimum distance $d = 4$. Corresponding generator matrix and trellis diagram are shown in Fig 1.

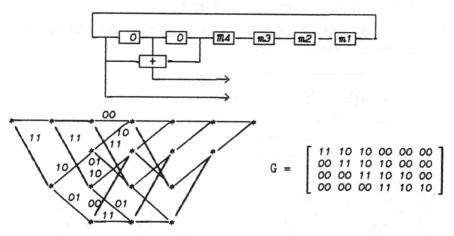

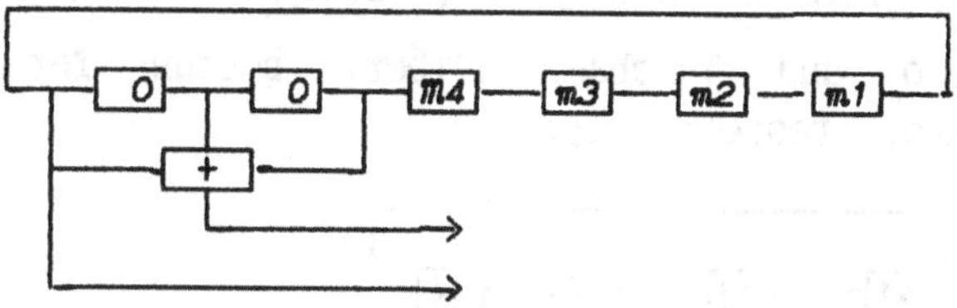

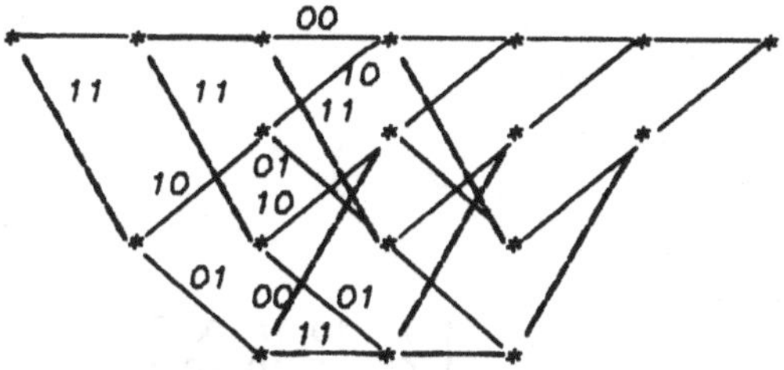

$$G = \begin{bmatrix} 11 & 10 & 10 & 00 & 00 & 00 \\ 00 & 11 & 10 & 10 & 00 & 00 \\ 00 & 00 & 11 & 10 & 10 & 00 \\ 00 & 00 & 00 & 11 & 10 & 10 \end{bmatrix}$$

Fig.1 Terminated code.

One can see that different code symbols are dependent on different number of message bits. It seems possible to shorten the code without loss of minimum distance by "bending" of the tail of the last columns of matrix G. The only difference in the encoder operating is that the encoder register is initialized with ν last binary messages. The resultant code is a code of length $n = kn_0$ which is called a tail biting code. Its encoder, the generator matrix and the trellis diagram are shown in Fig.2. We have linear $(8,4)$-code, $d = 4$, which is the Hamming code.

3. Minimum distance.

Consider a (n,k)-code obtained from a convolutional code with constraint length ν. Than ν is less than both the description of the code and the decoding are simpler. It is clear that if ν is large enough, we can construct codes with good minimum distance. On the contrary, if ν is small, then the minimum distance is very likely to be small too. The question is how small may be ν for the resultant block code to satisfy Varshamov-Gilbert bound. The following theorem gives the asymptotical answer.

Theorem 1. The tail biting code of length n, rate R, minimum distance $d = n\delta_V(R)$ does exist under the condition

$$\nu/n \geqslant \delta_V(R)/\delta_C(R),$$

where

$$\delta_C(R) = -R/\log_2(2^{(1-R)}-1);$$

$$\delta_V(R) = H^{-1}(1-R), \quad H(x) = -x\log_2 x - (1-x)\log_2(1-x),$$

are the Costello and Varshamov-Gilbert bounds for block and convolutional codes, respectively.

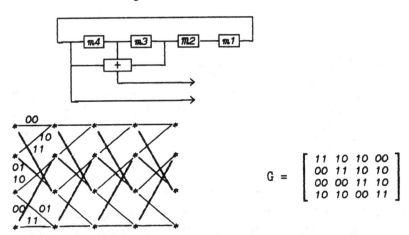

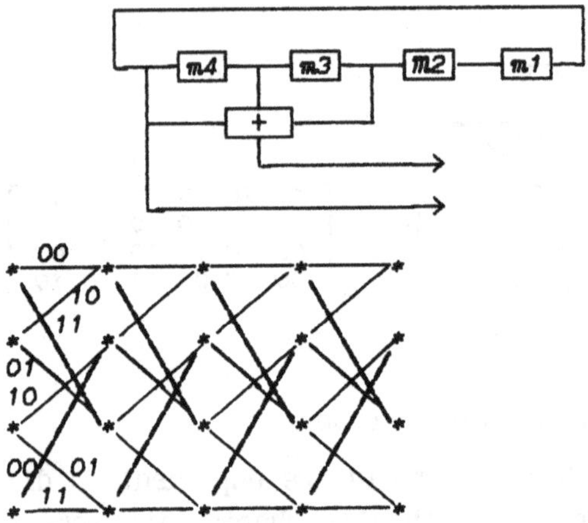

$$G = \begin{bmatrix} 11 & 10 & 10 & 00 \\ 00 & 11 & 10 & 10 \\ 00 & 00 & 11 & 10 \\ 10 & 10 & 00 & 11 \end{bmatrix}$$

Fig.2. Tail biting code.

This theorem has been proved in [9] by random coding arguments.

Not only asymptotical distance performances of tail biting codes are good. Almost all best linear codes of rate 1/2 and 1/3 were constructed as tail biting codes by computer search. A full table for found codes is presented in [9]. Some examples are in Table 1.

Table 1

Examples of the tail biting codes

rate	n,k,d	ν	code generators
1/2	8,4,4	2	4,7
	16,8,5	3	3,11
	18,9,6	3	7,13
	22,11,7	5	44,57
	24,12,8	6	51,127
	38,19,8	5	15,57
	40,20,9	6	131,117
	42,21,10	7	133,217

rate	n,k,d	ν	code generators
1/3	9,3,4	1	2,2,3
	15,5,7	3	1,7,13
	33,11,11	5	13,17,53
	45,15,14	7	221,333,373
	60,20,16	7	125,337,362
1/4	48,12,16	4	25,27,33,37
	52,13,19	6	134,135,147 165

In this table conventional octal notations for code generators have been used. For example, the first code of the table is the code shown in Fig.2.

4. Decoding.

Consider the encoder presented in Fig.2 and its trellis diagram. In this trellis some sequence of 8 binary digits corresponds to each path from beginning level to end level. Every node of the diagram corresponds to some state of the encoder register. Since the initial encoder state coincides with the final state, only those paths are corresponding to code words which start and end in the nodes with the same numbers.

To apply Viterbi decoding algorithm it is necessary to know the initial node. A straightforward procedure for maximum likelihood (ML) decoding of the tail biting code is to run Viterbi algorithm through the trellis 2^{ν} times, each time with a different initial node. After each trial we get the best path from the fixed initial node to corresponding end node. Choosing most likelihood path among 2^{ν} finding paths we get maximum likelihood decision. For this decoding algorithm the complexity is proportional to $n2^{2\nu}$ operations for one code word. The data given in the Table show that this algorithm for soft decoding is simpler than exhaustive search through all 2^k code words .

For further decoding simplification we have to find a method for identifying the initial encoder state (initial trellis node). To describe another decoding algorithm we introduce a special trellis called periodic trellis. This trellis represents M times repeated trellis of the code under consideration.

Decoder puts the metrics of all initial nodes equal to 0 and makes the Viterbi algorithm trial to find the path of length Mk branches closest to M times repeated received sequence. After that decoder "checks" all nodes of the best path. Consider the checking of the node I located at level τ. Decoder puts the metric of node I at level τ equal to 0 and puts metrics of all other nodes at level τ equal to $(-\infty)$. Viterbi algorithm runs from level τ to level $(\tau+k)$ and finds the best path to node I at level $(\tau+k)$ (cyclic path) and keeps the metric of the chosen path. After checking all nodes decoder chooses the best cyclic path. Left cyclic shift of the best cyclic path on $(\tau \bmod k)$ positions is the decoder decision.

We call this decoder an almost ML decoder. It makes the correct decision if it finds path of length Mk branches which has at least one intersection with the correct path. The decoder complexity is proportional to $Mn^2 2^{\nu}$. It has been proved in [10] that if M is proportional to ν then error probability is approximately the same as that of ML decoding.

5. Asymptotical error probability-decoding complexity function.

Let $X = \{x\}$ be the input alphabet and $Y = \{y\}$ be the output alphabet of a discrete memoryless channel and $\{ p(y/x) \}$ be the crossover probabilities, $\mathbf{p} = \{p(x)\}$ be the input probability distribution. Denote Gallager function as

$$E_0(\rho) = \max_{\mathbf{p}} - \log \sum_y \Big[\sum_x p(x)p(y/x)^{1/(1+\rho)}\Big]^{(1+\rho)}.$$

We denote Gallager's and Viterbi's random coding exponents for block and convolutional codes as

$$E_G(R) = \max_{\rho \in [0,1]} E_0(\rho) - \rho R ,$$

$$E_V(R) = E_0(\rho(R)), \quad \rho(R) = \min \{ 1, \rho_0(R)\} \tag{1}$$

respectively, where $\rho_0(R)$ is the non-negative solution of the equation

$$E_0(\rho) = \rho R .$$

Employing these denotations we can write the following theorem which establishes the error probability bounds for tail biting codes.

Theorem 2. There exists a tail biting code of length n, rate $1/n_0$, constraint length $\nu \geqslant nE_G(R)/E_V(R)$ with error probability of ML decoding

$$P_{eML} \leqslant \exp\{-n [E_G(R) - o(n)]\},$$

and error probability of AML decoding

$$P_{eAML} \leqslant \exp\{-n [E_G(R) - o(n) - o(M)]\},$$

where $o(z) \to 0$ if $z \to \infty$ and parameter M is the number of periods of periodic trellis.

It follows from Theorem 2 that if ν and M are large enough then codes with optimal error probability and decoding complexity æ about 2^{ν} do exist.

Corollary. There exist a block code and a decoding algorithm with complexity æ such that error probability P_e satisfies

$$P_e \sim æ^{-\rho(R)},$$

where $\rho(R)$ is defined by (1).

Note, that it is the same error probability-decoding complexity function as for Vierbi decoding convolutional codes.

6. Asymptotical code rate-decoding complexity function.

Define the asymptotical complexity coefficient as

$$\gamma(R) = \lim_{n \to \infty} (\log \varkappa)/n ,$$

where $\varkappa$ is the decoding complexity for a code of length n and rate R. We consider only block codes with minimum distance $d = \delta_V(R)$, i.e. codes satisfying asymptotical Varshamov-Gilbert bound. The known bounds for $\gamma(R)$ when soft decisions are used by decoder are the following

a) $\gamma(R) = R$ for the conventional ML decoding;

b) $\gamma_W(R) = 1 - R$ for the Wolf algorithm [7];

c) $\gamma_D(R) = \begin{cases} R \quad , R \leqslant 1 - H(0.25), \\ H(2\delta_V(R)) - H(\delta_V(R)), \ 1 - H(0.25) \leqslant R \leqslant 1; \end{cases}$

for the concatenated codes decoding by Dumer algorithm [8].

It can be proved by exploiting Theorems 1 and 2 that for the ML decoding of tail biting codes

$$\gamma_{ML}(R) = \gamma_D(R)$$

and for the AML decoding

$$\gamma_{AML}(R) = R\ \delta_V(R)/\ \delta_C(R).$$

Functions $\gamma(R)$, $\gamma_W(R)$, $\gamma_{ML}(R)$, $\gamma_{AML}(R)$ are shown in Fig.3.

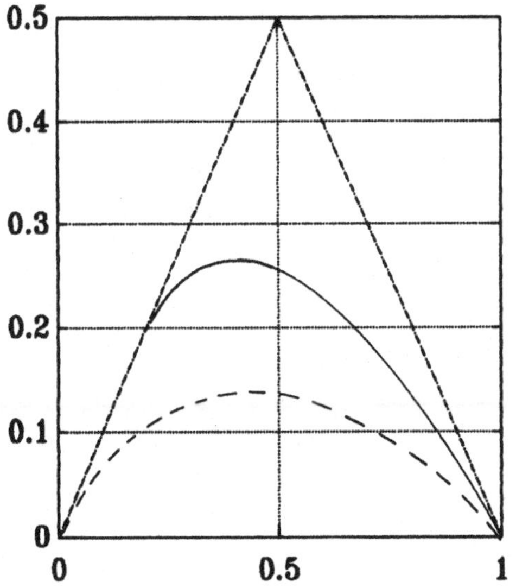

–·–·–·– $\gamma(R)$ and $\gamma_W(R)$, ——— $\gamma_{ML}(R)$, - - - - - $\gamma_{AML}(R)$

Fig.3. Asymptotical code rate-decoding complexity functions.

References

[1] H.Ma, J.K.Wolf. On the tail biting convolutional codes. IEEE Trans. Commun., 1986, v.34, 2, p.104-111.

[2] G.D.Forney. Convolutional codes, 2: Maximum likelihood decoding. Information and Control, 1974, v.25, 3, p. 222-266.

[3] K.Sh.Zigangirov, V.D.Kolesnik. Centered codes. Probl. Pered. Inf., 1984, v.20, 2, p.3-17.

[4] G.S.Evseev. On the complexity of decoding linear codes. Probl. Pered. Inf., 1983, v.19, 1, p.3-8.

[5] A.M.Barg, I.I.Dumer. An algorithm of concatenative decoding with incomplete code vectors look through. Probl. Pered. Inf., 1986, v.22, 1, p.3-10.

[6] Krouck E.A. A bound on the decoding complexity of linear block codes. Probl. Pered. Inf., 1989, v.25, 3, p.103-106.

[7] J.K.Wolf. Efficient maximum likelihood decoding of linear block codes using a trellis. IEEE Trans. Inform. Theory, 1978, v.24, 1, p.76-80.

[8] I.I.Dumer. On complexity of maximum likelihood decoding for the best concatenated codes. 8-th conf. on coding and inf. theory, Moscow-Kuybyshev, 1981, 2, p.66-69.

[9] Kudryashov B.D., Zakharova T.G. Block codes from convolutional codes. Probl. Pered. Inf.,1989, v.25, 4, p.98-102.

[10] Kudryashov B.D. Decoding of block codes obtained from convolutional codes.Probl.Pered. Inf., 1990, v.26, 3, p.18-26.

PARTIAL ORDERING OF ERROR PATTERNS FOR MAXIMUM LIKELIHOOD SOFT DECODING

Jakov Snyders

Department of Electrical Engineering - Systems, Tel Aviv University, Ramat Aviv 69978, Israel

Abstract

The error patterns, encountered in maximum likelihood soft decision syndrome decoding of a binary linear block code, can be partially ordered in a way that *only* the minimal elements have to be scored. The ordering requires a usually short sorting proceedure applied to the confidence values of the hard-detected bits. In this paper some properties of the minimal elements are derived. We also present bounds on the number of minimal elements with particularly large Hamming weight.

1. Introduction

The error patterns encountered in maximum likelihood soft decision syndrome decoding of a binary linear block code are usually rather numerous. They may be partially ordered in a way that only the minimal elements, whose number for most codes is a small fraction of the total number of error patterns, are to be scored for achieving maximum likelihood decoding. We shall present some properties of the minimal elements. Such properties can be useful for *a*) forming a set that includes the set of minimal elements, thereby facilitating maximum likelihood decoding, as well as for *b*) construction of a subset of minimal elements, for obtaining suboptimal decoders with reduced error rate.

Let C be an (n,k,d) binary linear code of blocklengh n, dimension k and minimum distance d. Denote $m = n-k$, the number of check bits of C. Let $H = (h_1, h_2, \cdots, h_n)$ be a parity-check matrix of C. Assume, without essential loss of generality, that $d \geq 3$. (In case that $d < 3$, implementation of the general decoding algorithm considered here will obviously start by converting the problem to one with $d \geq 3$ as follows: *a*) the all-0 columns of H are punctured, and *b*) a single column, out of each set of identical nonzero columns of H, is retained and labeled with weight which is the minimum weight over the set.) Let codewords with equal probability, 2^{-k}, be transmitted through a memoryless channel. Denote by v be the bit-by-bit hard-detected version of the received word, and denote by z the syndrome corresponding to v. Assume that $z \neq 0$. Let us assign to each h_i a *weight* $\rho(h_i)$, defined to be the magnitude of the log-likelihood ratio (often termed confidence value) of the associated bit v_i of v. Denote $\Gamma = \{h_i: i = 1,2, \cdots ,n\}$. The *weight* $\rho(\Phi)$ of $\Phi \subset \Gamma$ is defined to be the sum of the weights of the vectors belonging to Φ. A set of t linearly independent elements of Γ that adds up to z will be called a *t-pattern*. Thus $t \leq m$. A *pattern* is a t-pattern with unspecified, or otherwise indicated, cardinality. For avoiding trivial situations, we shall let $m \geq 3$.

An algorithm [1], [2] for carrying out maximum likelihood decoding may now be phrased as follows:

1) among all the t-patterns, with t ranging up to m, find one with least weight, then
2) complement v at the locations identified by the pattern that was found.

Due to the extremly large number of patterns (for most codes), implementation of step 1) has rarely [1]-[3] been addressed. Some nonexhaustive search procedures [4]-[6] are well known. In [7] a method was considered for substantially reducing the family of patterns to be scored, while maintaining optimality of decoding. A version of it now follows. Let y_j for j=1,2, $\cdots$,m denote certain columns of H that satisfy

$$\rho(y_j) = min\{\rho(h): \; h \in \Gamma - Ls(\{y_i: \; i=1,2, \cdots ,j-1\})\} \tag{1}$$

where Ls stands for linear span. Denote

$$\Omega_m = \{y_1, y_2, \cdots ,y_m\}.$$

Evidently, Ω_m is a basis for $GF(2)^m$. Let Δ and Φ be patterns. We shall write $\Delta \leq \Phi$ if

a) $\Delta - \Phi \subset \Omega_m$ and

b) there exists an injection ψ: $\Delta - \Phi \rightarrow \Phi - \Delta$ such that for each $y_j \in \Delta - \Phi$ we have $\psi(y_j) \in \Gamma - Ls(y_1, y_2, \cdots ,y_{j-1})$.

The relation $\leq$ thus defined is clearly reflexive and transitive. It is also antisymmetric in view of the following obvious statement.

Lemma 1: If Δ and Φ are patterns such that $(\Delta - \Phi) \cup (\Phi - \Delta) \subset \Omega_m$ then $\Delta = \Phi$.

Thus a partial ordering of patterns has been established. An element Σ is called Ω_m-minimal, or *minimal* for short, if $\Theta \leq \Sigma$ implies $\Theta = \Sigma$. For example, assuming that $\Phi = \{y_2, \; y_1+y_2+y_3+y_5, \; y_4+y_5+y_6, \; y_5\}$ is a pattern, we have $\Delta := \{y_2, \; y_1+y_2+y_3+y_5, \; y_4, \; y_6\} < \Phi$. Also, $\Delta^* := \{y_1, \; y_3, \; y_4+y_5+y_6\} < \Phi$. It is easily verified that both Δ and Δ^* are minimal elements. Clearly, only the minimal elements need to be scored for accomplishing maximum likelihood decoding. (Minimal elements are called Ω_m-survivors in [7]).

2. A simple example

In this Section let C be the (7,4,3) code. Assuming that $z = y_1$, the only minimal element is $\{y_1\}$. Similarly, if $z = y_3$ then only $\{y_3\}$ is minimal. At the other extreme, i.e. if $z = y_1+y_2+y_3$, then the minimal patterns are $\{z\}$, $\{y_1, y_2+y_3\}$, $\{y_2, y_1+y_3\}$, $\{y_3, y_1+y_2\}$ and $\{y_1, y_2, y_3\}$. Thus there is at most one minimal pattern with cardinality three, whereas there are (for any $z \neq 0$) four 3-patterns.

Aided with the knowledge of minimal patterns, decoding of C is performable by a total of 11 - 13 real number operations (additions and comparisons). The complexity of another possible scheme of implementation is 6 - 17. The former scheme achieves the smallest known worst-case complexity, namely 13. In fact, it coincides with the decoder of [2]. However, in contrast to C, for most codes only a rather small fraction of patterns is minimal. Consequently, use of minimal patterns may considerably increase the efficiency of decoding of some codes. In particular, for longer Hamming codes a substantial computational gain was achieved by scoring minimal patterns [7].

Of course, the set of minimal patterns for C includes the set of minimal patterns for *any* code with $m = 3$ (with respect to an identical value of z).

3. Some properties of minimal patterns

As already observed, the family of minimal elements depends on the value of the syndrome z. More generally, we have the following

Lemma 2: There exists a unique pattern Δ such that $\Delta \subset \Omega_m$. Δ is minimal. With the possible exception of Δ, no pattern Φ with $|\Phi| \geq r$ is minimal, where $r = max\{i: y_i \in \Delta\}$.

Proof: The existence follows by $z \in Ls(\Omega_m)$. The uniqueness and minimality are implied by Lemma 1. The proof is completed with the aid of (1).

Lemma 3: Let $\Phi = \{h_i: i \in L\}$, where $|L| \geq 3$, be a pattern. If for a proper subset K of L with $|K| \geq 2$ and some $y_j \in \Omega_m$

$$\sum_{i \in K} h_i = y_j \tag{2}$$

then Φ is not minimal.

Proof: Set $\Delta = (\Phi \cup \{y_j\}) - \{h_i: i \in K\}$. Then Δ is a pattern and $|\Delta| < |\Phi|$. Furthermore, $\rho(h_i) \geq \rho(y_j)$ for some $i \in K$, as otherwise $h_i \in Ls(\Omega_{j-1})$ for all $i \in K$ by (1), in contradiction of (2). Hence $\Delta < \Phi$.

Our main result now follows.

Theorem 4: Let an $(m-j)$-pattern Φ be minimal. Then $|\Phi \cap \Omega_m| \geq m-2j$.

Proof: For $j = 0$ the result is implied by Lemma 2. Let $j \geq 1$. The statement is straightforwardly verifiable for $m = 3$ (see Section 2). Assume its validity for some $m \geq 3$ (and all j). We shall proceed by induction. Suppose that a pattern Φ, $\Phi \subset GF(2)^{m+1}$, with cardinality $(m+1)-j$ is Ω_{m+1}-minimal but $|\Phi \cap \Omega_{m+1}| < m+1-2j$. Without loss of generality, we identify $Ls\{y_1, y_2, \cdots, y_m\}$ with $GF(2)^m$. Consequently $\{y_1, y_2, \cdots, y_m\} = \Omega_m$. Now form Φ^* as follows: expand all vectors of Φ in terms of Ω_{m+1}, then delete y_{m+1} from all the expansions. Two cases have to be considered.

Case *a*) $y_{m+1} \notin \Phi$. Then $|\Phi^*| = m+1-j = m-(j-1)$ and $|\Phi^* \cap \Omega_m| = |\Phi \cap \Omega_{m+1}| < m+1-2j$. However, Φ^* is Ω_m-minimal, hence by the assumption we should have $|\Phi^* \cap \Omega_m| \geq m-2(j-1) = m+2-2j$.

Case *b*) $y_{m+1} \in \Phi$. Then $|\Phi^*| = m-j$ and, since according to Lemma 3 no vector in Φ is equal to $y_{m+1}+y_i$ for any $i \leq m$, we conclude that $|\Phi^* \cap \Omega_m| = |\Phi \cap \Omega_{m+1}| - 1 < m-2j$. As Φ^* is Ω_m-minimal, the inequality contradicts the assumption.

By Theorem 4 many, quite possibly the vast majority, of patterns with cardinality exceeding $m/2$ are non-minimal. Many additional patterns, even those with small cardinality, may shown to be non-minimal by other means, e.g. Lemma 3.

Let $N(t)$ be the number of Ω_m-minimal t-patterns. By Theorem 4 $N(m) \leq 1$, with equality iff Ω_m is a pattern.

Let δ_{ij} be the Kronecker delta.

Corollary 5: Assume that Ω_m is a pattern. Then an $(m-1)$-pattern Φ is minimal iff $|\Omega_m - \Phi| = 2$. Furthermore,

$$N(m-1) \leq \binom{m}{2} \delta_{d3} \tag{3}$$

with equality for $d = 3$ iff $y_i+y_j \in \Gamma$ for all $i < j \leq m$.

Only the Hamming codes meet the condition $y_i+y_j \in \Gamma$, $i < j \leq m$, with respect to all values of z.

Theorem 6: Let $\Phi' = \Omega_m - \{y_{j_1}, y_{j_2}, \cdots, y_{j_s}\}$, where $1 \leq j_1 < j_2 < \cdots < j_s < m$, be a pattern. Φ' is a minimal $(m-1)$-pattern, provided $s = 1$. If $j_s + 1 < m$ then an $(m-1)$-pattern Φ, $\Phi \neq \Phi'$, is minimal iff it includes $\Psi := \{y_1, y_2, \cdots, y_{j_s}\}$ and contains all but precisely two members of $\Omega_m - \Psi$. If $j_s+1 = m$ then no pattern with cardinality $m-1$ is minimal, exclusive of Φ' in case that $s = 1$.

Proof: Φ' is minimal since $\Phi' \subset \Omega_m$. Denote $R = \{j_1, j_2, \cdots, j_s\}$. Considering the case $j_s+1 < m$ and $s \geq 2$, let Φ satisfy the conditions of the Theorem. $\Phi' < \Phi$ can not hold because $\Phi' - \Phi = \{y_a, y_b\}$ for some $b > a > j_s$ and $\Phi - \Phi' = \{y_i\colon i \in R\} \cup \{g\}$ where $g = y_a + y_b + \sum_{i \in R} y_i$. Suppose now that $\Delta \leq \Phi$ for some pattern Δ, $\Delta \neq \Phi'$. Then $\Delta - \Phi \subset \Omega_m$. Since $\Phi - \{g\} \subset \Omega_m$, this implies $g \in \Delta$, but then $\Phi - \Delta \subset \Omega_m$, thus yielding $\Delta = \Phi$ in view of Lemma 1. This concludes the proof of sufficiency of the conditions. The necessity of the conditions, as well as the conclusions for the case $j_s+1 = m$, are implied by Theorem 4, Lemma 3 and the following Lemma 7.

Lemma 7: Let an Ω_m-minimal Φ contain

$$h = y_a + y_b + \sum_{j \in T} y_j$$

where y_a and y_b are distinct members of $\Omega_m - \Phi$ and $\{y_j\colon j \in T\} \subset \Omega_m \cap \Phi$ (T is possibly empty). Then $max\{j\colon y_j \in T\} < min\{a, b\}$.

Proof: Assume that $max\{j\colon y_j \in T\} > min\{a,b\}$ and set $\Delta = (\Phi \cup \{y_a, y_b\}) - \{h\} - \{y_i\colon i \in T\}$. Then $\Delta < \Phi$.

Corollary 8: If the assumption of Theorem 6 prevails then

$$N(m-1) \leq \delta_{s1} + \binom{m-j_s}{2} u_{s-d+3} \tag{4}$$

where $u_i = 1$ for $i \geq 0$ and $u_i = 0$ otherwise. Furthermore, the bound is attained iff

$$\sum_{i=1}^{s} y_{j_i} + \sum_{i \in T} y_i \in \Gamma$$

for each $T \subset \{j_s+1, j_s+2, \cdots, m\}$ with $|T| = 2$.

Regarding the determination of the least weighing minimal pattern with cardinality $m-1$, a large j_s represents a twofold advantage: the minimal $(m-1)$-patterns, whose number is confined by (4) to be small, have an intersection with cardinality of at least j_s. At the other extreme, if $j_s < d-3$, then $N(m-1) \leq 1$ according to (4).

The following overall bound is implied by (3) and (4) for the case that y_m participates in the expansion of z:

$$N(m-1) \leq \delta_{d4} + \binom{m-d+3}{2}.$$

By Lemma 2 for the remaining case, i.e. $z \in Ls\{y_1, y_2, \cdots, y_{m-1}\}$, we have $N(m-1) \leq 1$, with equality iff $\{y_1, y_2, \cdots, y_{m-1}\}$ is a pattern.

References

[1] H. Miyakawa and T. Kaneko, "Decoding algorithm of error-correcting codes by use of analog weights," *Electronics Communications in Japan*, vol. 58-A, pp. 18-27, 1975.

[2] J. Snyders and Y. Be'ery, "Maximum likelihood soft decoding a binary block codes and decoders for the Golay codes," *IEEE Trans. Inform. Theory*, vol. IT-35, *pp.* 963-975, 1989.

[3] S. Litsyn and E. Nemirovsky, "Simplification of maximum likelihood decoding of block codes on the basis of the Viterbi algorithm," *Trudy Instituta Ingenerov Radio*, vol. 2, pp. 17-27, 1988 (in Russian).

[4] G.D. Forney, Jr., *Concatenated Codes.* Cambridge, Massachusetts: The MIT Press, *pp.* 61-62, 1966.

[5] D. Chase, "*A* class of algorithms for decoding block codes with channel measurement information," *IEEE Trans. Inform. Theory*, vol. IT-18, *pp.* 170-182, 1972.

[6] E.R. Berlekamp, "The construction of fast, high-rate, soft decision block decoders," *IEEE Trans. Inform. Theory*, vol. IT-29, pp.372-377, 1983.

[7] J. Snyders, "Reduced lists of error patterns for maximum likelihood soft decoding," *IEEE Trans. Inform. Theory*, to be published.

A FAST MATRIX DECODING ALGORITHM FOR RANK-ERROR-CORRECTING CODES

Ernst M. Gabidulin

Moscow Institute of Physics and Technology
Institutskii per., 9,
141700 DOLGOPRUDNYI, Moscow Region, USSR
e-mail: gab@ippi.msk.su

Abstract. The so-called term-rank and rank metrics and appropriate codes were introduced and investigated in [1 - 7]. These metrics and codes can be used for correcting array errors in a set of parallel channels, for scrambling in channels with burst errors, as basic codes in McEliece public key cryptosystem [8], *etc*. For codes with maximal rank distance (MRD codes) there exists a fast decoding algorithm based on Euclid's Division Algorithm in some non-commutative ring [6]. In this paper a new construction of MRD codes is given and a new fast matrix decoding algorithm is proposed which generalizes Peterson's algorithm [9] for BCH codes.

1. Backgrounds

Consider a rectangular $N \times n$ matrix with entries from the field $GF(q)$:

$$\mathbf{E} = \begin{Vmatrix} e_{11} & \dots & e_{1n} \\ \cdot & \dots & \cdot \\ e_{N1} & \dots & e_{Nn} \end{Vmatrix}. \quad (1)$$

Definition 1. The *Term Rank Norm* $N_{TR}(\mathbf{E})$ of a matrix $\mathbf{E}$ is defined as the *minimal number* of lines (rows or columns) containing all non-zero elements of this matrix. The *Term Rank Distance* $d_{TR}(\mathbf{E}_1,\mathbf{E}_2)$ between matrices $\mathbf{E}_1$ and $\mathbf{E}_2$ is defined as the term rank norm of their difference: $d_{TR}(\mathbf{E}_1,\mathbf{E}_2) = N_{TR}(\mathbf{E}_1 - \mathbf{E}_2)$.

Example. The term rank of a matrix

$$\begin{Vmatrix} 1 & 1 & 1 & 1 \\ 0 & 1 & 0 & 0 \\ 0 & 1 & 0 & 0 \\ 0 & 1 & 0 & 0 \end{Vmatrix}$$

is equal to 2 because all non-zero elements are in the first row and in the second column and it is impossible to choose a smaller number of lines containing all non-zero elements.

An s by r array error is defined as a matrix containing all non-zero elements in s rows and m columns. The rank metric is matched with array errors in the sense that if a code has a code term rank distance d it can correct all array errors if $s+m < d/2$. This metric and appropriate codes were introduced and investigated in [1 - 4, 7].

Definition 2. The *Rank Norm* $N_R(\mathbf{E})$ of a matrix $\mathbf{E}$ is defined as the rank of $\mathbf{E}$ over $GF(q)$: $N_R(\mathbf{E}) = r(\mathbf{E};q)$ (all axioms of the norm are satisfied for the rank function). The *Rank Distance* $d_R(\mathbf{E}_1,\mathbf{E}_2)$ between matrices $\mathbf{E}_1$ and $\mathbf{E}_2$ is defined as the rank of their difference: $d_R(\mathbf{E}_1,\mathbf{E}_2) = r(\mathbf{E}_1-\mathbf{E}_2;q)$.

Note that for any matrix $\mathbf{E}$

$$N_H(\mathbf{E}) \geqslant N_{TR}(\mathbf{E}) \geqslant N_R(\mathbf{E}),$$

where $N_H(\mathbf{E})$ is the Hamming norm of $\mathbf{E}$. This means that any code correcting, say, t-fold rank errors can also correct t-fold term rank errors (or array errors) or t-fold Hamming (random) errors.

Let us give another definition of the rank norm. Consider the columns of the matrix $\mathbf{E}$ in (1) as components of a representation of some elements of a "large" field $GF(q^N)$ over the fixed basis $\omega_1, \omega_2, \ldots, \omega_N$:

$$e_j = \sum_{i=1}^{N} e_{ij}\omega_i, \quad j = 1, \ldots, n, \tag{2}$$

and introduce n-vector $\mathbf{e} = (e_1,e_2,\ldots,e_n)$ over $GF(q^N)$ instead of a $N\times n$ matrix $\mathbf{E}$ over $GF(q)$.

Definition 3. The *Rank Norm* $N_R(\mathbf{E})$ of a vector $\mathbf{e}$ over $GF(q^N)$ is defined as a *maximal number* of components of this vector which are linearly independent over $GF(q^N)$. The *Rank Distance* between $\mathbf{e}_1$ and $\mathbf{e}_2$ is the rank of their difference: $d_R(\mathbf{e}_1,\mathbf{e}_2) = r(\mathbf{e}_1-\mathbf{e}_2;q)$.

We consider codes consisting of vectors over a "large" field

$GF(q^N)$. Note that we can always reformulate all the results for matrices over $GF(q)$.

Without loss of generality we can consider the case $N = n$.

Consider a linear (n,k)-code and let d be code distance for some metric. For all mentioned metrics - Hamming, Term Rank, and Rank - the following Singleton-type bound is true:

$$R = k/n \leqslant 1 - (d-1)/n \tag{3}$$

(for the Hamming metric it is a well known result, but for other metrics it holds because the Hamming norm is not less than the term rank- or rank norm for any vector). If in (3) equality holds, then codes are called MDS codes for the Hamming metric, and MRD codes (Maximum Rank Distance) for other metrics.

A general class of MRD codes was proposed for the first time in [5, 6]. Parity-check matrix H of a MRD code with code distance d is defined as follows

$$H = \begin{Vmatrix} h_1 & h_2 & \dots & h_n \\ h_1^{[1]} & h_2^{[1]} & \dots & h_n^{[1]} \\ \cdot & \cdot & \dots & \cdot \\ h_1^{[d-2]} & h_2^{[d-2]} & \dots & h_n^{[d-2]} \end{Vmatrix}, \tag{4}$$

where here and hereafter "$[i]$" in an exponent is shorthand for "q^i" and where elements $h_i \in GF(q^n)$, $i = 1, \dots, n$, must be linearly independent over $GF(q)$. There exists a fast decoding algorithm for these codes based on Euclid's division algorithm in some non-commutative ring (see [6] for details).

In Section 2 we describe a new class of MRD codes.

In Section 3 we propose a new fast decoding algorithm with the parity-check matrix given by (4). This algorithm can be considered as a generalization of Peterson's matrix algorithm [9] for BCH codes.

2. A New Class of MRD Codes

Choose k elements $\alpha_1, \alpha_2, \ldots, \alpha_k$ and $r = n-k$ elements $\beta_1, \beta_2, \ldots, \beta_r$ in such a way that all n elements $\alpha_1, \alpha_2, \ldots, \alpha_k, \beta_1, \beta_2, \ldots, \beta_r$ are linearly independent over $GF(q)$ (in other words, they form a basis of $GF(q^n)$ over $GF(q)$).

Introduce k linearized polynomials $F_s(z)$, $s = 1, 2, \ldots, k$, of degree q^{k-1}, where $F_s(z)$ has elements $\alpha_1, \alpha_2, \ldots, \alpha_{s-1}, \alpha_{s+1}, \ldots, \alpha_k$ and all their linear combinations with coefficients from $GF(q)$ as roots.

Form generator matrix as follows

$$G = \begin{Vmatrix} F_1(\alpha_1) & 0 & 0 & \ldots & 0 & F_1(\beta_1) & F_1(\beta_2) & \ldots & F_1(\beta_r) \\ 0 & F_2(\alpha_2) & 0 & \ldots & 0 & F_2(\beta_1) & F_2(\beta_2) & \ldots & F_2(\beta_r) \\ \cdot & \cdot & \cdot & \ldots & \cdot & \cdot & \cdot & \ldots & \cdot \\ 0 & 0 & 0 & \ldots & F_k(\alpha_k) & F_k(\beta_1) & F_k(\beta_2) & \ldots & F_k(\beta_r) \end{Vmatrix}. \tag{5}$$

Theorem 1. A code with generator matrix given by (5) is a MRD code of rate $R = k/n$ and code rank distance $d = n-k+1$.

Proof. Consider any code vector

$$\mathbf{g} = \mathbf{u}\,G = (u_1, u_2, \ldots, u_k)G =$$
$$(u_1F_1(\alpha_1), u_2F_2(\alpha_2), \ldots, u_kF_k(\alpha_k), \sum_{s=1}^{k} u_sF_s(\beta_1), \ldots, \sum_{s=1}^{k} u_sF_s(\beta_r)) \tag{6}$$

and calculate how many linear combinations of coordinates of this vector with coefficients from $GF(q)$ can be equal to zero. Let $\mathbf{v} = (v_1, \ldots, v_n)$, $v_i \in GF(q)$, be one of these combinations. Then

$$0 = \mathbf{g}\mathbf{v}^T = \sum_{i=1}^{k} v_i \sum_{s=1}^{k} u_sF_s(\alpha_i) + \sum_{i=1}^{r} v_{i+k} \sum_{s=1}^{k} u_sF_s(\beta_i) =$$
$$\sum_{s=1}^{k} u_sF_s(\sum_{i=1}^{k} v_i\alpha_i + \sum_{i=1}^{k} v_{i+k}\beta_i). \tag{7}$$

The linearized polynomial $\Phi(z) = \sum_{s=1}^{k} u_sF_s(z)$ of degree not greater than q^{k-1} can not have more than $k-1$ roots which are linearly independent over $GF(q)$. Thus, there are at least $d = r+1 = n-k+1$

coordinates of the vector $\mathbf{g}$ which are non-zero and linearly independent over $GF(q)$, and the rank of $\mathbf{g}$ is equal at least to d. ∎

3. A New Fast Decoding Algorithm of MRD Codes

Consider a MRD code with parity-check matrix given by (4). Assume that $\mathbf{g} = (g_1, \ldots, g_n)$ is a code vector; $\mathbf{e} = (e_1, \ldots, e_n)$ is an error vector, and $\mathbf{y} = \mathbf{g} + \mathbf{e}$ is a received vector.

Decoding starts with calculating the syndrome vector

$$\mathbf{s} = (s_1, s_2, \ldots, s_{d-2}) = \mathbf{y}\mathbf{H}^T = \mathbf{e}\mathbf{H}^T. \tag{8}$$

One must determine the error vector on the basis of the known syndrome vector $\mathbf{s}$. Assume that the rank norm of the error vector $\mathbf{e}$ is m. Then it can be written in the form

$$\mathbf{e} = \mathbf{E}\mathbf{Y} = (E_1, \ldots, E_m)\mathbf{Y}, \tag{9}$$

where $E_1, \ldots, E_m$ are linearly independent over $GF(q)$, while $\mathbf{Y} = (Y_{ij})$ is an $m \times n$ matrix of rank m with entries from $GF(q)$. Then instead of (8) we can write

$$\mathbf{s} = \mathbf{E}\mathbf{Y}\mathbf{H}^T = \mathbf{E}\mathbf{X}, \tag{10}$$

where the matrix $\mathbf{X} = \mathbf{Y}\mathbf{H}^T$ has the form

$$\mathbf{X} = \begin{Vmatrix} x_1 & x_1^{[1]} & \ldots & x_1^{[d-2]} \\ x_2 & x_2^{[1]} & \ldots & x_2^{[d-2]} \\ \cdot & \cdot & \ldots & \cdot \\ x_m & x_m^{[1]} & \ldots & x_m^{[d-2]} \end{Vmatrix}, \tag{11}$$

and elements

$$x_p = \sum_{j=1}^{n} Y_{ij} h_j, \quad p = 1, \ldots, m \tag{12}$$

are linearly independent over $GF(q)$. Equation (10) is equivalent to the following system of equations in the unknowns $E_1, \ldots, E_m, x_1, \ldots, x_m$:

$$\sum_{i=1}^{m} E_i x_i^{[p]} = s_p, \quad p = 0, 1, \ldots, d-2. \tag{13}$$

Assume that a solution of this system has been found. Then we can determine the matrix Y from system (12) and the error vector **e** from (9). Thus the decoding problem reduces to finding a solution of system (13) for the smallest possible value of m. We describe the matrix algorithm for a solution of this system.

We introduce the linearized polynomial

$$\sigma(z) = \sum_{i=0}^{m} \sigma_i z^{[i]}, \quad \sigma_m = 1, \tag{14}$$

which has all linear combinations with coefficients from $GF(q)$ of linearly independent unknowns x_i's from (13) as roots. Unknowns $\{\sigma_i\}$ and coordinates $\{s_i\}$ of the syndrome vector are related by means of linear equations as the following theorem shows.

Theorem 2. For any $i \geqslant 0$

$$s_{i+m} + \sigma_{m-1}^{[i]} s_{i+m-1} + \sigma_{m-1}^{[i]} s_{i+m-2} + \ldots + \sigma_0^{[i]} s_i = 0,$$
$$i = 0, 1, \ldots, d-2-m, \tag{15}$$

or, equivalently,

$$s_{i+m}^{[-i]} + \sigma_{m-1} s_{i+m-1}^{[-i]} + \sigma_{m-1} s_{i+m-2}^{[-i]} + \ldots + \sigma_0 s_i^{[-i]} = 0,$$
$$i = 0, 1, \ldots, d-2-m, \tag{16}$$

Proof. Multiply the ith equation of (13) by $\sigma_0^{[i]}$, the $(i+1)$th - by $\sigma_1^{[i]}$, ..., the $(i+m-1)$th - by $\sigma_{m-1}^{[i]}$, the $(i+m)$th - by $\sigma_m^{[i]} = 1$ and sum up:

$$\sum_{s=0}^{m} E_s \sum_{k=0}^{m} \sigma_k^{[i]} x_s^{[k+i]} = s_{i+m} + \sigma_{m-1}^{[i]} s_{i+m-1} + \sigma_{m-1}^{[i]} s_{i+m-2} + \ldots + \sigma_0^{[i]} s_i. \tag{17}$$

Note that

$$\sum_{k=0}^{m} \sigma_k^{[i]} x_s^{[k+i]} = \Big(\sum_{k=0}^{m} \sigma_k x_s^{[k]}\Big)^{[i]} = (\sigma(x_s))^{[i]} = 0$$

by definition of the polynomial $\sigma(z)$. Thus, equation (15) is true.

The extraction of the root of degree $[i] = q^i$ from both sides of

(15) gives equation (16). ■

It is easy to obtain a new system

$$(\sigma_0, \sigma_1, \ldots, \sigma_m)\mathbf{M}_m = -(s_m, s_{m+1}^{[-1]}, \ldots, s_{2m-1}^{[-m+1]}), \quad (18)$$

from system (16), where

$$\mathbf{M}_i = \begin{Vmatrix} s_0 & s_1^{[-1]} & \ldots & s_{i-1}^{[-i+1]} \\ s_1 & s_2^{[-1]} & \ldots & s_i^{[-i+1]} \\ . & . & \ldots & . \\ s_{i-1} & s_i^{[-1]} & \ldots & s_{2i-2}^{[-i+1]} \end{Vmatrix} . \quad (19)$$

What about the solvability of this system? The answer is given by the following Lemma.

Lemma. If the rank norm of the error vector is $m \leqslant (d-1)/2$, then the matrix $\mathbf{M}_i$ is non-singular if $i = m$ and is singular if $i > m$.

Proof. Replace s_i in (19) by their expressions from (13) and reduce the matrix $\mathbf{M}_i$ to the following form

$$\mathbf{M}_i = \begin{Vmatrix} x_1 & x_2 & \ldots & x_i \\ x_1^{[1]} & x_2^{[1]} & \ldots & x_i^{[1]} \\ . & . & & . \\ x_1^{[i-1]} & x_2^{[i-1]} & \ldots & x_i^{[i-1]} \end{Vmatrix} \begin{Vmatrix} E_1 & E_1^{[-1]} & \ldots & E_1^{[-i+1]} \\ E_2 & E_2^{[-1]} & \ldots & E_2^{[-i+1]} \\ . & . & \ldots & . \\ E_i & E_i^{[-1]} & \ldots & E_i^{[-i+1]} \end{Vmatrix} . \quad (20)$$

We must use $x_i = 0$ and $E_i = 0$ if $i > m$. If $i = m$ then both matrices in the right side of (20) are non-singular because both $x_1, \ldots, x_m$ and $E_1, \ldots, E_m$ are linearly independent over $GF(q)$. If $i > m$ both matrices in the right side of (20) are singular because they contain zero-columns and zero-rows. ■

Theorem 2 and the Lemma give us the following procedure for decoding:

- calculate the syndrome vector **s**;
- calculate $\det(\mathbf{M}_t)$, where $t = (d-1)/2$; if $\det(\mathbf{M}_t) \neq 0$, then solve (18), find all t linearly independent roots $x_1, \ldots, x_t$ of the equation $\sigma(z) = 0$; find $E_1, \ldots, E_t$ from (13); find Y_{ij} from (12); find the error vector e from (9);
- if $\det(\mathbf{M}_t) = 0$, then calculate $\det(\mathbf{M}_{t-1})$, ..., and so on.

Using this algorithm we have to solve systems of linear equations a few times.

It is interesting to note that the matrix algorithm can be modified for decoding the MDR codes proposed in Section 2.

4. References

1. Korzhik V.I., Gabidulin E.M. *A Class of Two-Dimensional Codes Correcting Lattice-Pattern Errors*. -In: The 2nd International Symposium on Information Theory, September 2-8, 1971, Tsahkadzor, Armenian SSR, Abstracts of papers, pp. 114-116.
2. Gabidulin E.M., Korzhik V.I. *Codes Correcting Lattice-Pattern Errors*. Izvestija vuz'ov, Radioelektronika, 1972, Vol. 15, No. 4, pp. 492-498 (in Russian).
3. Gabidulin E.M., Sidorenko V.R. *Codes for a Set of Parallel Channels*. -In: The 3rd International Symposium on Information Theory, Abstracts of papers, P. 2. -Moscow - Tallinn, 1973, pp.54-57.
4. Gabidulin E.M., Sidorenko V.R. *Codes for some memory channels*. -In: The 4th International Symposium on Information Theory, Abstracts of papers, P. 2. Moscow - Leningrad, pp. 24-28.
5. Gabidulin E.M. *Optimal Array-Error-Correction Codes*. -In: The 6th International Symposium on Information Theory, Abstracts of papers, P. 2. Moscow - Tashkent, pp. 75-77.
6. Gabidulin E.M. *Theory of Codes with Maximum Rank Distance*. -Problems of Information Transmission (Проблемы передачи информации - Problemy peredachi informatsii). Translated from Russian. *C/B* Consultants Bureau, New York, July 1985, pp. 1-12. Russian original Vol. 21, No. 1, January-March, 1985.
7. Gabidulin E.M. *Optimal Codes Correcting Array Errors*. -Problemy peredachi informatsii, 1985, vol. 21, No. 2 (in Russian).
8. Gabidulin E.M., Paramonov A.V., and Tretjakov O.V. *Ideals over a Non-Commutative Ring and their Application in Cryptology*. -Proceedings of Eurocrypt'91.
9. Peterson W.W., Weldon E.J., Jr. *Error-Correcting Codes*. Second Edition, MIT Press, Cambridge, Mass. 1972.

A FAST SEARCH FOR THE MAXIMUM ELEMENT OF THE FOURIER SPECTRUM

Alexey E. Ashikhmin
Institute for Problems of Information Transmission,
Ermolovoy 19, Moscow USSR

Simon N. Litsyn
Department of Electrical Engineering - Systems, Tel-Aviv University
Ramat-Aviv, 69978 ISRAEL
Email: litsyn @ genius.tau.ac.il

ABSTRACT

An algorithm for determining the index of the maximum spectrum element of the Fourier transform is proposed. Its complexity is linear in the length of the processed sequence. The algorithm leads to the correct decision if the Euclidean distance between the sequence and a basis vector does not exceed $\sqrt{(q/2)}$, where q is the length of the sequence.

INTRODUCTION

The problem of searching for the maximum element of the Fourier spectrum attracts wide attention because of its importance in a number of practical problems in communication and control. The obvious algorithm consists in sequential implementing of the FFT and searching for the maximum among the spectral components. In [1] an approach was proposed for a faster implementation of the search of the maximum spectral element (MSE) for the Walsh transform . The algorithm of [1] exploits an idea of truncating intermediate results of the fast Walsh transformation algorithm. The complexity of the aforementioned algorithm is linear in the length of the processed sequence. The procedure was essentially based on the Kronecker product representation of the Walsh functions. Unfortunately, for the case of the discrete Fourier base this does not lead to a direct generalization of the above algorithm.

In the following paper we give a description of an algorithm for a fast implementation of the search of the MSE. If the maximal factor of the length of the sequence is not growing with the length then we get a linear order of complexity in the length for the algorithm. Furthermore, we show that it gives the correct decision if the Euclidean distance between the sequence and a basis vector does not exceed $\sqrt{(q/2)}$, where q is the length of the sequence.

Let us introduce some notations. Let $\omega_q = \exp(2\pi j/q)$, q be an arbitrary positive integer; The Fourier base orthogonal vectors of length n=q have the shape

$$v(b) = (v_0, v_1, ..., v_{q-1}) = (\omega_q^{0*b}, \omega_q^{1*b}, ..., \omega_q^{(q-1)*b}),$$
$$b=0, 1,..., q-1.$$

The orthogonality of the sequences is a corollary of the well-known property

$$(1) \qquad \sum_{k=0}^{q-1} \omega_q^{bk} = \begin{cases} 0, & \text{if } b\neq 0; \\ q, & \text{otherwise.} \end{cases}$$

Hence, (here * denotes complex conjugation)

$$|(v(b_1), v^*(b_2))| = | \sum_{k=0}^{q-1} \omega_q^{(b_2-b_1)k} | =$$

$$= \begin{cases} 0, & \text{if } b_1 \neq b_2; \\ q, & \text{if } b_1 = b_2. \end{cases}$$

In what follows we shall call the parameter b the vector b-number.

Consider the following problem. Let $\mathbf{w} = (w_0, w_1, \ldots, w_{q-1})$ be a sequence. Our goal is to find the index of the maximal modulus of a spectrum component, or ,equivalently, to determine the b-number of the base vector closest to **w** in Euclidean metric.

THE FFT BASED ALGORITHM

For the sake of completeness we present here a version of the FFT based algorithm. We consider the Cooley-Tukey FFT procedure [2]. We shall write it down in a form usefull for further development of a faster procedure.

Let $q=q_1q_2$ (here we use only two factors of q, the general case can be deduced recursively; that allows us to simplify the description of the algorithm).

ALGORITHM A.

1. Organize the matrix $G=[g_{i,j}]$, $i=0,1,\ldots,q_1-1$; $j=0,1,\ldots,q_2-1$, as follows:

$$G = \begin{matrix} w_0 & w_{q_1} & \cdots & w_{q_2q_1-q_1} \\ w_1 & w_{q_1+1} & \cdots & w_{q_2q_1-q_1+1} \\ \cdots & \cdots & \cdots & \cdots \\ \cdots & \cdots & \cdots & \cdots \\ w_{q_1-1} & w_{2q_1-1} & \cdots & w_{q_2q_1-1} \end{matrix}$$

2. Implement the discrete Fourier transformation of dimension q_1 over columns of the matrix G. We get a matrix $T=[t_{i,j}]$ with

$$t_{i,j} = \sum_{k=0}^{q_1-1} \omega_{q_1}^{ik} g_{k,j} \, ,$$

$$i=0,1,\ldots,q_1-1; \quad j=0,1,\ldots,q_2-1.$$

3. Compute the matrix $R=[r_{i,j}]$, where

$$r_{i,j} = \omega_q^{ij} t_{i,j} \, ,$$

$$i=0,1,\ldots,q_1-1; \quad j=0,1,\ldots,q_2-1.$$

4. Compute the matrix $F=[f_{k,i}]$,

$$f_{k,i}=\sum_{j=0}^{q_2-1} \omega_{q_2}^{ij} r_{k,j},$$

$$i=0,1,\ldots,q_1-1;\quad j=0,1,\ldots,q_2-1.$$

5. Pick the vector of spectrum coefficients $s=[s_j]$, where

$$s_j = f_{q_1 i + k},$$

$$j=0,1,\ldots,q-1.$$

6. Compute $\mathbf{h}=[h_j]$, $h_j = s_j s_j^*$, $j=0,1,\ldots,q-1$.
7. Compute b= index of the maximal element in **h**.
8. End.

The above algorithm requires $q(q_1+q_2+1)$ complex additions, $q(q_1+q_2-1)$ complex multiplications and $(q-1)$ comparisons. Recurrent use of the algorithm for factors of q_1 and q_2 leads to further simplifications of the algorithm. The minimal complexity of the algorithm has an order of $q \log q$.

The algorithm guarantees the correct result in the case when the Euclidean distance between the processed vector and one of the base vectors does not exceed $\sqrt{(q/2)}$. It is a consequence of the orthogonality property of the Fourier base and the triangle inequality (see [3] for details).

THE TRUNCATED ALGORITHM

Further development of the algorithm may be realized on the base of truncation on intermediate steps. The idea is in erasing all rows but the one with the maximal expected impact to the final result.

Let again $w=[w_i]$ be the considered vector, $i=0,1,\ldots,q-1$; $q=q_1 q_2 \ldots q_m$. Denote by n_i the product $q_{i+1}\ldots q_m$, $i=0,1,\ldots,m-1$; with evidently $q=n_0$. Assume $n_m=1$.

ALGORITHM B.

1. For I=1 to m do steps 2-6.
2. Organize matrix $G=[g_{i,j}]$, $i=0,1,\ldots,q_I-1$; $j=0,1,\ldots,n_I-1$,

$$G=\begin{matrix} g_{0,0} & g_{0,1} & \ldots & g_{0,n_I-1} \\ g_{1,0} & g_{1,1} & \ldots & g_{1,n_I-1} \\ \ldots & \ldots & \ldots & \ldots \\ \ldots & \ldots & \ldots & \ldots \\ g_{q_I-1,0} & g_{q_I-1,1} & \ldots & g_{q_I-1,n_I-1} \end{matrix} =$$

$$= \begin{bmatrix} w_0 & w_1 & \cdots & w_{n_I-1} \\ w_{n_I} & w_{n_I+1} & \cdots & w_{2n_I-1} \\ \cdots & \cdots & \cdots & \cdots \\ \cdots & \cdots & \cdots & \cdots \\ w_{(q_I-1)n_I} & w_{(q_I-1)n_I+1} & \cdots & w_{q_I n_I-1} \end{bmatrix} .$$

3. Implement discrete Fourier transformation of dimension q_I over columns of matrix G.We get the matrix $T=[t_{i,j}]$ with

$$t_{i,j}=\sum_{k=0}^{q_I-1} \omega_{q_I}^{ik} \, g_{k,j} \quad ,$$

$$i=0,1,...,q_I-1; \quad j=0,1,...,n_I-1.$$

4. Compute the vector $\mathbf{p}=[p_i]$, $i=0,1,...,q_I-1$;

$$p_i = \sum_{j=0}^{n_I-1} t_{i,j} t_{i,j}^* = \| t_i \|^2.$$

5. Compute b_I = index of the maximal element in **p**. If $I=m$ go to step 7.
6. Compute the vector $w=[w_j]$, $j=0,1,...,n_I-1$; where

$$w_j = \omega_q^{b_I j} \, t_{b_I,j} \; .$$

7. Let

$$b = \sum_{i=1}^{m} b_i \, n_i \quad .$$

8. The end.

The algorithm requires

$$\sum_{i=1}^{m} (2q_i+1) \, n_i$$

complex multiplications,

$$\sum_{i=1}^{m} ((q_i - 1)\, n_i + (n_i - 1) q_i)$$

complex additions, and

$$\sum_{i=1}^{m} (q_i - 1)$$

comparisons of real numbers.

Particularly, in the case $q = 2^m$ the following amount of operations is necessary: $(5q - 5)$ complex multiplications, $(3q - 2\log_2 q - 3)$ complex additions and $\log_2 q$ comparisons, i.e. the complexity is linear in q.

The complexity of Algorithm B changes with the order of the factors of q (in contrast to algorithm A). Direct verification shows that in order to minimize complexity the most preferable strategy is to sort factors of q in increasing order.

CONDITIONS FOR THE CORRECT DECISION

In what follows we'll derive conditions for the algorithm B to give the correct result.It depends essentially on the Euclidean distance between the considered vector and one of the base vectors.

Let

$$r_I = \sum_{i=I}^{m} b_i\, n_i\,.$$

Assume $w = (w_0, w_1, \dots, w_{q-1})$, $w_i = \omega_q^{-bi}$, $i=0,1,\dots,q-1$; i.e. w coincides with one of the base vectors. In this case according to the algorithm on the first step ($I = 1$) we have

$$g_{i,j} = \omega_q^{-bj}\, \omega_q^{-bn_1 i} = \omega_q^{-bj}\, \omega_{q1}^{-bi} = \omega_q^{-bj}\, \omega_{q1}^{-(r_1 q_1 + b_1) i} = \omega_q^{-bj}\, \omega_{q1}^{-b_1 i}\,.$$

Then,

$$t_{i,j} = \omega_q^{-bj} \sum_{k=0}^{q_1 - 1} \omega_{q1}^{-b_1 i}\, \omega_{q1}^{ik} = \omega_q^{-bj} \sum_{k=0}^{q_1 - 1} \omega_{q1}^{(i - b_1) k} =$$

$$= \begin{cases} 0, \text{ if } i \neq b_1 \\ w_q^{-bj} q_1, \text{ otherwise.} \end{cases}$$

Therefore,

$$p_i = \sum_{j=0}^{n_1-1} t_{i,j} t_{i,j}^* = \begin{cases} 0, \text{ if } i \neq b_1 \\ q_1 q, \text{ otherwise.} \end{cases}$$

Hence on the second step (I = 2) $w=[w_i]$, $i = 0, 1, \ldots, n_2-1$;

$$w_i = q_1 \, \omega_q^{-(b-b_1)i} = q_1 \, \omega_{n_1}^{-r_1 i}$$

is processed. By induction one may show that on the I-th step the initial vector $w=[w_i]$, $i=0,1,\ldots \ldots,n_I - 1$; is determined by expression

$$w_i = q_1 \ldots q_{I-1} \, \omega_{n_{I-1}}^{-r_{I-1} i} ,$$

and therefore ,

$$p_i = \begin{cases} 0, \text{ if } i \neq b_I \\ q_1 q_2 \ldots q_I q, \text{ otherwise.} \end{cases}$$

Hence, we have shown that the algorithm gives the correct result for the base vectors. Moreover, the large difference between values of p_i for wrong and correct values enables us to determine the correct value even for vectors far apart from the base ones.

W.l.o.g. we may assume $w_i = 1+ \xi_i$, $i=0,1,\ldots,q - 1$; where ξ_i are complex numbers. We are looking for conditions yielding $b_j = 0$, $j=0,1,\ldots, m$. Let us denote $\xi = (\xi_1, \ldots \ldots, \xi_{q-1})$.

Then on the I-th step provided correct decodings up to the (I - 1) -th step, we have

$$t_{0,j} = q_1 q_2 \ldots q_I +$$

$$= \sum_{k_I=0}^{q_I-1} \sum_{k_{I-1}=0}^{q_{I-1}-1} \sum_{k_{I-2}=0}^{q_{I-2}-1} \cdots \sum_{k_1=0}^{q_1-1} \xi_{k1\,n1 + k2n2 + \ldots + kI\,nI + j},$$

$$t_{i,j} = \sum_{k_I=0}^{q_I-1} \sum_{k_{I-1}=0}^{q_{I-1}-1} \sum_{k_{I-2}=0}^{q_{I-2}-1} \cdots \sum_{k_1=0}^{q_1-1} \xi_{k1\,n1 + k2n2 + \ldots + kI\,nI + j}\ \omega_{q_I}^{k_I i}$$

for $i \neq 0$.

$$p_0 = \sum_{j=0}^{n_I-1} t_{0,j} t_{0,j}^* ,$$

$$p_i = \sum_{j=0}^{n_I-1} t_{i,j} t_{i,j}^* .$$

Let $\phi_i(I, \xi) = p_0 - p_i$, $i \neq 0$. Using the method of Lagrange multipliers we may show (see [3] for details) that under the assumption

$$\| \xi \|^2 < q/2$$

the following inequality holds

$$\min\ \phi_i(I, \xi) > 0,$$

thus yielding the correct result within the Euclidean sphere of radius $\sqrt{(q/2)}$ centered on a base vector the base vectors. Moreover, from the proof of the mentioned result it follows that in step 4 of the algorithm an arbitrary positive power of the norm of a row of T may be used.

REFERENCES

1.S.Litsyn and O.Shekhovtsov, *Fast decoding algorithm for first order Reed-Muller codes*, Problems of Information Transmission, v.19, 2, pp.3-7, 1983.

2.R.Blahut, *Fast algorithms for digital signal processing*, Addison-Wesley P.C., 1985.

3.A.Ashikhmin and S.Litsyn, *Fast decoding of nonbinary orthogonal codes*, in preparation.

Coding Theorem for Discrete Memoryless Channels with Given Decision Rule.

Vladimir B. Balakirsky (USSR).

Consulting center "Method", Antonenko, 6a, 190107 Leningrad
&& "Lenpromstroiproekt", Leninsky, 160, 196158 Leningrad.

The maximal transmission rate over binary-input discrete memoryless channel is calculated when the decoding decision function is fixed in any additive manner.

Introduction.

As well-known, channel capacity was defined by C.Shannon [1] as the maximal rate for transmission over this channel with arbitrary small error probability. An optimization over all possible block codes and all possible decoding rules is produced during its calculating and the last optimum is achieved when maximum-likelihood decision function is maximized during decoding procedure. We consider a similar problem when the channel is defined by input alphabet X, output alphabet Y and transition probabilities $W = \{ W_x(y) \}$, when a block code G with rate R and length n consisting of codewords $\mathbf{x}_1,\dots,\mathbf{x}_M$ where $M = exp\{ nR \}$ is used for transmission. However, the decision function at the channel output is fixed in an additive manner, i.e. the function $d(x, y)$ is defined for all $x \in X$, $y \in Y$, and the decoder calculates $d(\mathbf{x}_1, \mathbf{y}),\dots, d(\mathbf{x}_M, \mathbf{y})$ where

$$d(\mathbf{x}, \mathbf{y}) = \sum_{j=1}^{n} d(x_j, y_j) \quad \text{for all } \mathbf{x} \in X^n$$

The estimation $\hat{m}$ of the transmitted message m is formed according to the following rule :

$$d(\mathbf{x}_{\hat{m}}, \mathbf{y}) = \min_{j=1,M} d(\mathbf{x}_j, \mathbf{y})$$

If $\hat{m} \neq m$ then a decoding error takes place. Let

$$P_e(G) = \max_{m=1,M} \sum_{\mathbf{y}} W_{\mathbf{x}_m}(\mathbf{y})\cdot\chi\left\{ \hat{m} \neq m \right\}$$

be the maximal decoding error probability; $W_{\mathbf{x}_m}(\ \mathbf{y}\ )$ is the conditional probability of the received word being $\mathbf{y}$ when $\mathbf{x}_m$ is transmitted, and $\chi\{\cdot\}$ is the indicator function of the event in braces.

Our purpose is to calculate the maximal transmission rate $C_d(\ W\ )$, defined by : if $R > C_d(\ W\ )$ then there exists $\varepsilon > 0$ such that $P_e(\ G\ ) > \varepsilon$ for any G; otherwise, if $R < C_d(\ W\ )$ then for all $\varepsilon > 0$ there exist n and a block code G such that $P_e(\ G\ ) < \varepsilon$. The considered problem is closely related to some open problems in information theory connected with arbitrary varying channels analysis and computation of channel capacity when decoding error probability is equal to zero [2]. Furthermore, this task may be interesting for applications of coding theory results when maximum-likelihood decoding is not used [3].

Results.

Let $P = \{\ P_x\ \}$ be some probability distribution on X and let PW and $d(\ P,W\ )$ be probability distribution on Y and average value of d, respectively, calculated in the ensemble $\{\ XY,\ P_x \cdot W_x(\ y\ )\ \}$, i.e.

$$PW(\ y\ ) = \sum_x P_x \cdot W_x(\ y\ )$$

$$d(\ P,W\ ) = \sum_{x,y} P_x \cdot W_x(\ y\ ) \cdot d(\ x,y\ )$$

Let $V = \{\ V_x(\ y\ )\ \}$ be some channel, and PV and $d(\ P,V\ )$ be defined similarly for the ensemble $\{\ XY,\ P_x \cdot V_x(\ y\ )\ \}$. Let

$$\overset{*}{C}_d(\ W\ ) = \max_P \overset{*}{C}_d(\ P,W\ ), \qquad (1)$$

$$\overset{*}{C}_d(\ P,W\ ) = \min_{\substack{V:\ PV = PW,\\ d(\ P,V\ ) \leqslant d(\ P,W\ )}} I(\ P,V\ ) \qquad (2)$$

where

$$I(\ P,V\ ) = H(\ PV\ ) - H(\ V \mid P\ ),$$

$$H(\ PV\ ) = -\sum_y PV(\ y\ ) \cdot \ln PV(\ y\ ),$$

$$H(\ V \mid P\ ) = -\sum_{x,y} P_x \cdot V_x(\ y\ ) \cdot \ln V_x(\ y\ ).$$

Theorem 1 (*direct coding theorem*).

$$C_d(W) \geq C_d^*(W)$$

for any W and d.

Theorem 2 (*converse*).

For a binary-input memoryless channel ($X = \{ 0,1 \}$) the inequality

$$C_d(W) \leq C_d^*(W)$$

is valid for all W and d.

Corollary.

$$C_d(W) = C_d^*(W)$$

for any binary-input memoryless channel W and any d.

Discussion.

1). If $d(x,y) = - \ln W_x(y)$ then the decoding rule coincides with maximum-likelihood decoding. Then the function $C_d^*(W)$ defined in (1) coincides with the channel capacity, and the converse theorem is valid for all memoryless channels.

2). The "universal" decoding rule when the decoder maximizes "mutual information" between codewords $\mathbf{x}_1,\ldots,\mathbf{x}_M$ and $\mathbf{y}$:

$$I(\mathbf{x}_{\hat{m}} , \mathbf{y}) = \max_{j = 1,M} I(\mathbf{x}_j , \mathbf{y}),$$

was investigated in [2,4]; here

$$I(\mathbf{x}, \mathbf{y}) = \sum_{x,y} n_{\mathbf{x},\mathbf{y}}(x,y) \cdot \ln \frac{n_{\mathbf{x},\mathbf{y}}(x,y)}{n_{\mathbf{x}}(x)}$$

and

$$n_{\mathbf{x},\mathbf{y}}(x,y) = \sum_{t = 1}^{n} \chi\left\{ x_t = x, y_t = y \right\},$$

$$n_{\mathbf{x}}(x) = \sum_{t = 1}^{n} \chi\left\{ x_t = x \right\},$$

As it was shown in [4], this decision function allows us to achieve the channel capacity $C(W)$ for any W : if P is defined by the equation $C(W) = I(P,W)$ and $R < C(W)$ then for any $\varepsilon > 0$ there exists "a fixed P-composition block code G " ($n_{\mathbf{x}_j}(x) = n \cdot P_x$ for all $x \in X$ and $j = 1, exp\{ n \cdot R \}$) such that $P_e(G) < \varepsilon$. Note, however, that $I(\mathbf{x}, \mathbf{y})$ is non-additive function and does not belong to the class of decision functions, which are considered in our paper.

3). Besides $P_e(G)$, the transmission reliability is characterized by average decoding error probability :

$$\overline{P_e(G)} = \frac{1}{M} \cdot \sum_{m = 1}^{M} \sum_{\mathbf{y}} W_{\mathbf{x}_m}(\mathbf{y}) \cdot \chi\left\{ \hat{m} \neq m \right\},$$

and the maximal transmission rate $\overline{C_d(W)}$ may be defined like $C_d(W)$. It is easy to see that

$$C_d(W) = \overline{C_d(W)} \qquad (3)$$

In fact, if $R < C_d(W)$, then for any $\varepsilon > 0$ there exists a code G with $2 \cdot exp\{ nR \}$ codewords such that $P_e(G) < \varepsilon$. The puncturing of $exp\{ nR \}$ codewords leads us to the code G' having $P_e(G') < \varepsilon$. Therefore, $C_d(W) \leqslant \overline{C_d(W)}$. However, $C_d(W) \geqslant \overline{C_d(W)}$ and (3) is proved.

To continue the discussion we need several definitions which are taken from [2].

A probability distribution P is "a type in X^n " if $n \cdot P_x$ are integers for all $x \in X$. Let $\mathcal{P}_n$ be the set of all possible types in X^n and $\mathbf{T}_P^n$ be the "set of sequences of type $P \in \mathcal{P}_n$" :

$$\mathbf{T}_P^n = \left\{ \mathbf{x} \in X^n : n_{\mathbf{x}}(x) = n \cdot P_x , x \in X \right\}$$

A conditional distribution V is "a conditional type in Y^n for a given $P \in \mathcal{P}_n$ " if $n \cdot P_x \cdot V_x(y)$ are integers for all $x \in X$, $y \in Y$. Let $\mathcal{P}_n(P)$ be the set of all conditional types in Y^n and

$$\mathbf{T}_V^n(\mathbf{x}) = \left\{ \mathbf{y} \in Y^n : n_{\mathbf{x},\mathbf{y}}(x,y) = n \cdot P_x \cdot V_x(y), \quad x \in X,\ y \in Y \right\}.$$

where $\mathbf{x} \in \mathbf{T}_P^n$.

4). Suppose that a block code G consists of codewords $\mathbf{x}_m \in \mathbf{T}_P^n$, $m = 1,M$ where $P \in \mathcal{P}_n$ and

$$R = \ln M / n > I(P,V) \qquad (4)$$

for some $V \in \mathcal{P}_n(P)$ satisfying (2). We are interested in the case

$$R < I(P,W), \qquad (5)$$

because, otherwise, the maximum-likelihood decoding does not achieve an arbitrary small error probability (we suppose that n is large enough and P maximizes the right side of (1)). Each codeword $\mathbf{x}_m$ may be interpreted as a point $\mathbf{x}_m \in \mathbf{T}_P^n$ in X^n, and each $\mathbf{y}$ as a point in Y^n. Let

$$\mathbf{A}_d^n(\mathbf{x}_m) = \Big\{ \mathbf{y} : d(\mathbf{x}_m, \mathbf{y}) \leqslant n \cdot d(P,W) \Big\}.$$

Suppose, $V,W \in \mathcal{P}_n(P)$. Then

$$\mathbf{T}_V^n(\mathbf{x}_m) \subseteq \mathbf{A}_d^n(\mathbf{x}_m), \quad \mathbf{T}_W^n(\mathbf{x}_m) \subseteq \mathbf{A}_d^n(\mathbf{x}_m).$$

Roughly speaking, one point from $\mathbf{T}_W^n(\mathbf{x}_m)$ will be realized after transmission $\mathbf{x}_m$ over the channel W. We can choose a code G for which "almost" all points from $\mathbf{T}_W^n(\mathbf{x}_m)$ do not coincide with the points from $\mathbf{T}_W^n(\mathbf{x}_j)$, $j \neq m$ (see (5)). However, to analyze the decoding error probability in our case it is necessary to estimate the number of d-"bad" points for $\mathbf{x}_m$, i.e. the number of common points between $\mathbf{T}_W^n(\mathbf{x}_m)$ and $\mathbf{A}_d^n(\mathbf{x}_j)$, $j \neq m$. As it is known [2],

$$| \mathbf{T}_V^n(\mathbf{x}_m) | \simeq exp\Big\{ n \cdot H(V \mid P) \Big\},$$

$$| \mathbf{T}_W^n(\mathbf{x}_m) | \simeq exp\Big\{ n \cdot H(W \mid P) \Big\},$$

and because V minimizes the right side of (2) it maximizes the conditional entropy function $H(V \mid P)$. Therefore, most points from $\mathbf{A}_d^n(\mathbf{x} \quad)$ have conditional type V. The size of intersection between $\mathbf{T}_V^n(\mathbf{x}_m)$ and the union of $\mathbf{T}_V^n(\mathbf{x}_j)$, $j \neq m$ is asymptotically the same as $| \mathbf{T}_V^n(\mathbf{x}_m) |$ (see (4)).In fact, the main result of the paper claims that for binary-input channels this condition is sufficient to obtain the size of intersection between $\mathbf{T}_W^n(\mathbf{x}_m)$ and the union of $\mathbf{A}_d^n(\mathbf{x}_j)$, $j \neq m$ asymptotically the same as $| \mathbf{T}_W^n(\mathbf{x}_m) |$.

5). Let $C_d(P,W)$ be the maximal transmission rate for fixed P-composition block codes (all codewords have type $P \in \mathcal{P}_n$). The number of types is the non-asymptotical function on n [2]; thus,

$$C_d(W) = \max_P C_d(P,W)$$

In fact, we estimate $C_d(P,W)$ by the function $C_d^*(P,W)$ defined in (2) during our analysis. We claim that

$$C_d(P,W) \geqslant C_d^*(P,W)$$

for all W and d (direct theorem) and

$$C_d(P,W) \leqslant C_d^*(P,W)$$

for all binary-input channels W and all d (converse). The statements presented before follow from these inequalities.

6). Let

$$d'(x,y) = d(x,y) + c_1(x) + c_2(y),$$

where $c_1(\cdot)$ and $c_2(\cdot)$ are any functions on X and Y. Then

$$C_d(P,W) = C_{d'}(P,W). \qquad (6)$$

In particular, if

$$c_1(x) = - \min_{y \in Y} \left\{ d(x,y) - d(0, y) \right\},$$

$$c_2(y) = - d(0, y),$$

then

$$d'(x,y) \geqslant 0, \quad d'(0,y) = 0 \text{ for all } y \in Y.$$

This approach may be used to simplify the operations.

7). It is easy to verify that the conditional probabilities V minimizing the right side of (2) may be expressed as

$$V_x(y) = f_x \cdot \varphi(y) \cdot exp\left\{ \alpha \cdot d(x,y) \right\}, \qquad (7)$$

where $\{ f_x \}$,$\{ \varphi(y) \}$ and parameter $\alpha \leqslant 0$ are chosen from the following conditions :

$$\sum_y V_x(y) = 1, \qquad (8)$$

$$\sum_x P_x \cdot V_x(y) = PW(y),$$

$$\sum_{x,y} P_x \cdot V_x(y) \cdot d(x,y) \leqslant d(P,W).$$

Hence, *minimizing the decision function d is maximum-likelihood decoding for the channel V when a fixed composition code is used.*

Examples.

Statement 1. Let

$$X = Y = \{ 0,1 \}, \quad \mathbf{P} = (P_0, P_1),$$

$$\mathbf{W} = \begin{bmatrix} 1 - \varepsilon_0 & \varepsilon_0 \\ \varepsilon_1 & 1 - \varepsilon_1 \end{bmatrix}, \qquad \mathbf{D} = \begin{bmatrix} c_0 & d_0 \\ d_1 & c_1 \end{bmatrix}.$$

Then

$$C_d(P,W) = C_d^*(P,W) = \tag{9}$$

$$= I(P,W)\cdot\chi\Big\{ \mathrm{sign}(1 - \varepsilon_0 - \varepsilon_1) = \mathrm{sign}(d_0 + d_1 - c_0 - c_1)\Big\},$$

where

$$\mathrm{sign}\ z = \chi\Big\{ z \geqslant 0 \Big\} - \chi\Big\{ z \leqslant 0 \Big\} \quad \text{for all } z.$$

Proof. Let

$$\mathbf{V} = \begin{bmatrix} 1 - \varepsilon_0' & \varepsilon_0' \\ \varepsilon_1' & 1 - \varepsilon_1' \end{bmatrix},$$

be the matrix containing transition probabilities V such that $C_d(P,W) = I(P,V)$ and let

$$\Delta_1 = \varepsilon_1 - \varepsilon_1'.$$

Because $PV = PW$, we write :

$$\varepsilon_0 - \varepsilon_0' = P_1\cdot\Delta_1 \ / \ P_0$$

and

$$H(V \mid P) = P_0\cdot\mathcal{H}(\varepsilon_0 - P_1\cdot\Delta_1 \ / \ P_0) +$$

$$+ P_1\cdot\mathcal{H}(\varepsilon_1 - \Delta_1)$$

where

$$\mathcal{H}(z) = -z\cdot \ln z - (1 - z)\cdot \ln(1- z).$$

The maximal value of $H(V \mid P)$ is equal to $H(PW)$. It is achieved if $\Delta_1 = \Delta_1^*$ where

$$\Delta_1^* = P_0\cdot(1 - \varepsilon_0 - \varepsilon_1).$$

Using (6) we conclude that $C_d(P,W)$ coincides with $C_{d'}(P,W)$ where the values of decision function d' are defined by the matrix :

$$\mathbf{D}' = \begin{bmatrix} 0 & 0 \\ d^* & c^* \end{bmatrix},$$

$$d^* = d_1 - c_0 - \min\Big\{ d_1 - c_0,\ c_1 - d_0 \Big\},$$

$$c^* = c_1 - d_0 - \min\Big\{ d_1 - c_0,\ c_1 - d_0 \Big\}.$$

Then

$$d'(P,V) = P_1 \cdot (\varepsilon_1' \cdot d^* + (1 - \varepsilon_1') \cdot c^*),$$

$$d'(P,W) = P_1 \cdot (\varepsilon_1 \cdot d^* + (1 - \varepsilon_1) \cdot c^*)$$

and either $d^* = 0$ or $c^* = 0$. If $d^* = 0$ then $\Delta_1 \leqslant 0$ (this follows from the inequality : $d'(P,V) \leqslant d'(P,W)$). Hence, $H(V \mid P)$ is maximized when $\Delta_1 = \Delta_1^*$ if $\Delta_1^* \leqslant 0$ and when $\Delta_1 = 0$ if $\Delta_1^* > 0$. Otherwise ($c^* = 0$), it is maximized when $\Delta_1 = \Delta_1^*$ if $\Delta_1^* \geqslant 0$ and when $\Delta_1 = 0$ if $\Delta_1^* < 0$. The union of considered cases leads us to (9).

Statement 2. Let

$$X = \{ 0,1 \}, \quad Y = \{ 0, 1, \ldots, L - 1 \},$$

$$P = (1/2, 1/2),$$

$$\mathbf{W} = \begin{bmatrix} \varepsilon_0 & \varepsilon_1 & \ldots\ldots \varepsilon_{L-1} \\ \varepsilon_{L-1} & \varepsilon_{L-2} & \ldots\ldots \varepsilon_0 \end{bmatrix},$$

$$\mathbf{D} = \begin{bmatrix} d_0 & d_1 & \ldots\ldots d_{L-1} \\ d_{L-1} & d_{L-2} & \ldots\ldots d_0 \end{bmatrix},$$

Then

$$C_{d}(, W) = C_d(P,W) = H(PW) - H(V \mid P),$$

where V is defined by the matrix

$$\mathbf{V} = \begin{bmatrix} \varepsilon_0' & \varepsilon_1' & \ldots\ldots \varepsilon_{L-1}' \\ \varepsilon_{L-1}' & \varepsilon_{L-2}' & \ldots\ldots \varepsilon_0' \end{bmatrix},$$

with

$$\varepsilon_j' = (\varepsilon_j + \varepsilon_{L-1-j}) \cdot \frac{exp\left\{ \alpha \cdot d_j \right\}}{exp\left\{ \alpha \cdot d_j \right\} + exp\left\{ \alpha \cdot d_{L-1-j} \right\}} \; ; \; j = 0,L-1$$

and parameter $\alpha \leqslant 0$ is chosen from the condition :

$$\sum_{j=0}^{L-1} \varepsilon_j' \cdot (d_j + d_{L-1-j}) = \sum_{j=0}^{L-1} \varepsilon_j \cdot (d_j + d_{L-1-j})$$

We use (7) and verify the validity of condition (8) with $f_0 = f_1 = 1$. This proves statement 2.

References.

1. *C.E.Shannon* A Mathematical Theory of Communication. *Bell Syst. Techn.J.*, 1948, **27**, p.379-423, 623-656.
2. *I.Csiszár, J.Körner* Information Theory. Coding Theorems for Discrete Memoryless Systems. *Academiai Kiado, Budapest*, 1981.
3. *G.C.Clark, J.B.Cain* Error-Correction Coding for Digital Communications. *Plenum Press, New York and London*, 1987.
4. *Goppa V.D.* Nonprobabilistic Mutual Information without Memory. - *Probl.Control Inform.Theory*, 1975, **4**, p.97-102.

DECODING FOR MULTIPLE-ACCESS CHANNELS

Irina E. Bocharova
Leningrad Aircraft Equipment Institute
Hertsen str. 67, 190000 Leningrad USSR

Abstract: We discuss some results in coding and decoding for multiple-access channels. A new approach to choosing codes and users' energies for that channel is described. For the case when a bit error probability and a decoding complexity are fixed we construct a region of achievable code rates.

1. Introduction

The two-user multiple-access channel is a model of a communications system where two independent messages from two sources are transmitted over a total channel simultaneously. Such multiple-access communications system is shown in figure 1. The multiple-access channel model that we use is a generalization of the binary adder channel with two inputs. In other words if $\mathbf{s}_{kj} \in S_k$, $k=1,2$ denotes the channel input vectors and $\mathbf{r}_j$ denotes the channel output vector, we assume that

$$\mathbf{r}_j = \mathbf{s}_{1j} + \mathbf{s}_{2j} + \mathbf{n}_j, \quad j=1,2,\ldots$$

where $\mathbf{n}_j$ denotes the q-dimensional Gaussian vector such that $\bar{\mathbf{n}}_j=0$ and $\overline{\mathbf{n}_j\mathbf{n}_i^T} = \delta_{ij}(N_0/2)\mathrm{I}$, δ_{ij} is the Kronecker symbol, I denotes the identity matrix. Input signals are vectors from $\mathbf{R}^q$ and each signal has only one nonzero component. It is equal to $\sqrt{E_k}$, $k=1,2$ for the k-th user's signal, where E_k denotes a signal energy. The signal set S_k for the k-th input consists of q orthogonal vectors. We shall consider the case of perfect block (node) and symbol synchronization between transmitters.

To encode a message sequence for the k-th source, the q-ary code $C_k(d_k,R_k)$ is used, where d_k denotes the free distance and R_k denotes the code rate in bits per symbol. We assume that a q-ary pseudorandom sequence $\{\eta_{kj}\}$ is added modulo-q to the encoded sequence $\{c_{kj}\}$. The resultant sequence is $\{\hat{c}_{kj}\}$.

There is a one-to-one correspondence between the symbols $\{\hat{c}_{kj}\}$ and vectors $\mathbf{s}_{kj} \in S_k$. A one-to-one mapping of the set $\{\hat{c}_{kj}\}$ to the signal set is arbitrary.

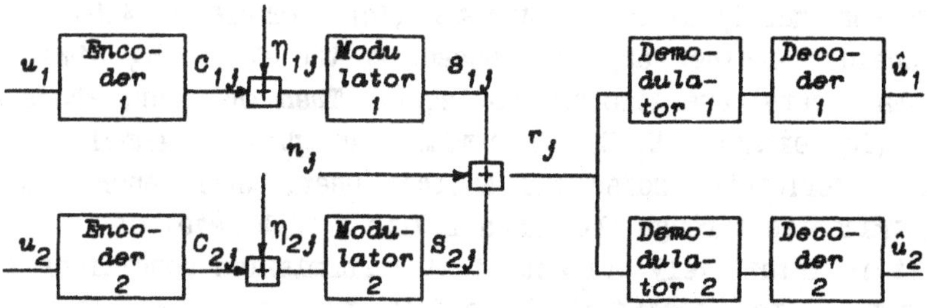

Fig.1 Model of the multiple-access communications system

The detailed classification of research areas for multiple-access channels is given in [1]. The reduced form of that classification is shown in figure 2.

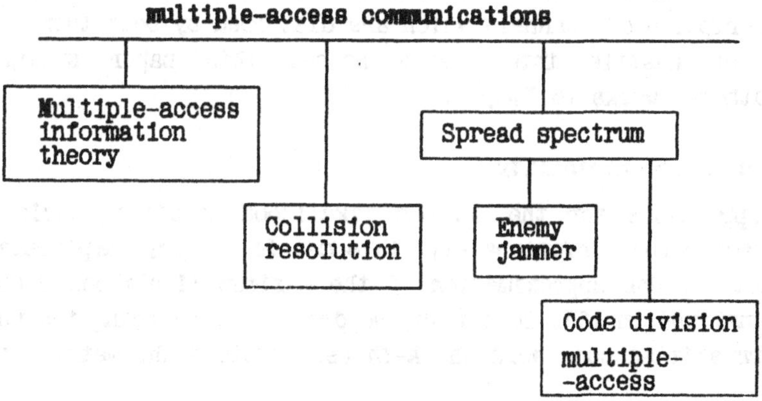

Fig.2 Classification of multiple-access systems

In the case under consideration, decisions are made by the individual decoders D_1 and D_2 . For the decoder D_k the encoded sequence $\{\hat{c}_{lj}\}$, $l \neq k$ can be considered as the other-user interference noise. According to the classification from [1] the above system corresponds to the traditional spread spectrum approach which treats everybody else's transmission as "jamming by the enemy". In other words the interference from other users is treated as additional noise. Note that code-division multiple-access systems [2,3] use knowledge of the users' waveforms enabling the receiver to demodulate the data streams of each user, upon observation of the sum of the transmitted signals, perturbed by additive noise. We assume the absence of such knowledge.

The conventional spread spectrum approach considers both users' energies to be equal $(E_1=E_2=E)$. The main problem is to construct special codes which enable one to divide transmitted sequences in noise or noiseless multiple-access systems (for example [1,4,5]). But we shall consider another approach based on the following simple idea. Assume that users' energies are not equal. Then the user, whose energy is large (for example, $E_1=E$), transmits through the channel with a low level of interference noise and another user, whose energy is small ($E_2<E$), transmits through the channel with a high level of interference noise. It is intuitively evident, that transmission conditions for the first user do not significantly differ from those for the channel without interference. As for the second user, the channel is not so bad as it seems. It is easy to detect the interference noise due to its high level and this noise can be eliminated with a high reliability.

Our approach is grounded on the using of the almost maximum likelihood decoding metric for such channel and optimization of the user's energy. Note that the channel model we use can be also exploited for the description of channels which are disturbed by both thermal and other-user or hostile interference noise. This paper summarizes previous author's works [6,7,8,9].

2. Bounds on error probability

The expressions for the maximum likelihood decoding metric have been obtained in [6]. Unfortunately, this metric is very complicated to use in practice. One approximation of the maximum likelihood decoding metric is analyzed in [6]. To obtain the decision concerning the number l of the transmitted code word the k-th user employs the metric in the form

$$\Gamma_k(l,\bar{\mathbf{r}},\alpha_k) = \sum_j \Gamma_k(l,\mathbf{r}_j,\alpha_k),$$

where

$$\Gamma_k(l,\mathbf{r}_j,\alpha_k) = \Gamma_k^{(1)}(l,\mathbf{r}_j) + \alpha_k \Gamma_k^{(2)}(l,\mathbf{r}_j),$$

α_k denotes the parameter of the optimization, $\bar{\mathbf{r}}= \mathbf{r}_1,\mathbf{r}_2, \ldots$.

We do not give exact formulas for functions $\Gamma_k^{(1)}$ and $\Gamma_k^{(2)}$, but note that $\Gamma_k^{(1)}$ is a correlation metric (this metric is optimum for the channel with additive white Gaussian noise). The metric $\Gamma_k^{(2)}$ is a nonlinear function of the signal and the observation $\mathbf{r}_j$, which takes into account the presence of the other-user interference.

It is shown in [6], that if N_0 tends to zero or q tends to infinity, the metric $\Gamma_k(\cdot,\cdot,\cdot)$ asymptotically coincides with the maximum likelihood metric. We shall call the metric Γ_k the generalized correlation metric.

In [6] the Viterbi decoding algorithm for convolutional codes using the metric Γ_k is exploited to obtain the almost maximum likelihood decoding. The bounds on the error probability for decoding using the metric Γ_k have been obtained in [6]. They are given in the following form

$$P_{Ek} \leqslant \min_{s>0,\alpha_k} B(s,d_k)T_k(D)|_{D=\hat{g}_k(s,\alpha_k)},$$

$$P_{Bk} \leqslant \min_{s>0,\alpha_k} B(s,d_k)(dT_k(D,I)/dI)|_{D=\hat{g}_k(s,\alpha_k),I=1},$$

where $B(s,d_k)=1/(2\sqrt{\pi d_k}s)$, s is a parameter, $T_k(D)$ is the transfer function of the code C_k and $\hat{g}_k(\cdot,\cdot)$ is the estimate for the generating function of the random variable

$$\Gamma_k(\mathbf{r}_j,m,\alpha_k) - \Gamma_k(\mathbf{r}_j,l,\alpha_k)\ ,\ l\neq m.$$

We do not give expressions for $\hat{g}_k(\cdot,\cdot)$ because they are rather complicated. Formulas for $\hat{g}_k(\cdot,\cdot)$ can be found in [6]. Note, that

$$\hat{g}_k(s,0)=\left[1+q^{-1}\left(2sh\left[s\sqrt{E_1E_2/2E_k}\right]\right)^2\right]exp\left\{-sE_k+s^2E_kN_0/2\right\}\ ,$$

i.e. $\hat{g}_k(s,0)$ coincides with the generating function corresponding to the respective correlation metric [10]. Let $\hat{g}_k = \min_{s\geqslant 0,\alpha_k} \hat{g}_k(s,\alpha_k)$. It can be shown that

$$\lim_{\sqrt{E_1E_2/E_k}\to\infty} \hat{g}_k = \lim_{q\to\infty} \hat{g}_k = \hat{g}_k\Big|_{\sqrt{E_1E_2/E_k}=0} = exp(-E_k/2N_0)$$

and if N_0 tends to zero $\hat{g}_k \sim exp\left[-min\left\{E_k,\left(\sqrt{E_1}-\sqrt{E_2}\right)^2\right\}/2N_0\right]$. In other words if the energy of the other-user interference noise tends to infinity or if the dimension of the signal space tends to infinity, $\hat{g}_k$ coincides with the estimate for the respective generating function of the correlation receiver in the absence of the other-user interference noise.

3. Two approaches to select users' energies

As shown in [6] there are two possible approaches to choosing the pair of energies (E_1,E_2) when we use $\Gamma_k(\cdot,\cdot,\alpha_k)$.

The first approach uses the maximum value of the bit error probability P_{Bk} as a criterion. We assume that codes C_1 and C_2 are fixed. Then one should choose the pair (E_1,E_2) to minimize the following value $\max_{k=1,2}\ \min_{s>0,\alpha_k} P_{Bk}$. If $C_1=C_2$ then one should minimize

$\max_{k=1,2} \hat{g}_k$. For example if E/N_0=7-13 dB and q=2 the optimum ratio E_2/E_1 is nearly 0.6.

The second approach uses pairs of achievable code rates (R_1,R_2) as a criterion. The error probability and the decoding complexity are assumed to be fixed. The user who has the maximum signal energy (for example $E_1=E$) is assumed to transmit by means of a high rate convolutional code and the second user who has the signal energy $E_2< E$ is assumed to transmit by means of a low rate code. Note that bounds on the sum of the achievable code rates, obtained in [6] are not asymptotical those ones. But they are useful in practice when we need to choose the pair of fixed codes (C_1,C_2) for the transmission.

4. Conventional codes for multiple-access channel

To realize the second approach we need *q-ary* convolutional codes which enable one to divide the transmitted sequences. But it is intuitively evident that, if we use the decoding metric Γ_k , at the output of the demodulator we obtain a code sequence almost free of the other-user interference noise. Hence, if we employ conventional convolutional codes (optimal for the channel with additive white Gaussian noise) in the above channel we may obtain some code gain.

What do we know about *q-ary* convolutional codes ? The detailed tables for binary convolutional codes (code rate $\leqslant$ 4/5) and their performances are given in [11]. Several classes of *q-ary* (q>2) codes are known (for example, the orthogonal [12], the semi-orthogonal [12] and the *dual-k* codes [13]). Unfortunately, there are not enough of the codes to construct the bounds on pairs of achievable code rates.

In [7] we have considered one class of high rate binary convolutional codes. These codes are not special codes for the multiple-access communications, but they are optimum in the sense of decoding complexity. Let us denote as (ν,d_f,R,q) convolutional codes with constraint length ν, free distance d_f, code rate R over the alphabet of size q. In [7] the codes $(m,3,(2^m-1)/2^m,2)$, m=1,2,... have been considered. These codes are defined by the semi-infinite generator matrix

$$G = \begin{Vmatrix} G_0 & G_1 & 0 & 0 & 0 & \dots 0 & 0 & \dots \\ 0 & G_0 & G_1 & 0 & 0 & \dots 0 & 0 & \dots \\ \dots & \dots & \dots & \dots & \dots & \dots & \dots & \dots \\ 0 & 0 & 0 & 0 & 0 & \dots G_0 & G_1 & \dots \\ \cdot & \cdot & \cdot & \cdot & \cdot & \cdot & \cdot & \dots \end{Vmatrix},$$

where all matrices are of size $(2^m-1)\times 2^m$,

$$G_0 = \left\| \begin{array}{c|c} \multicolumn{2}{c}{G_H^m} \\ \hline 0 & G^* \end{array} \right\| , \quad G_1 = \left\| \begin{array}{c|c} \multicolumn{2}{c}{0} \\ \hline 0 & I_m \end{array} \right\| , \quad G^* = \left\| \begin{array}{c|c} I_m & \begin{array}{c} 1 \\ 1 \\ \vdots \\ 1 \end{array} \end{array} \right\| .$$

The matrix G_H^m is the generator matrix of the $(2^m,2^m-m-1)$ Hamming code given in the systematic form, I_m is the identity matrix of size $(m\times m)$. The expression for the transfer function $T(D)$ for this class of codes is presented in [7]. As an example the region of the achievable code rates for the case, when codes with rates 1/6, 1/4, 1/3, 1/2, 2/3, 3/4, 4/5, 7/8 are assumed to be allowed, is shown in figure 3. Energies (E_1,E_2) are shown in parenthesis. Using the generalized correlation metric and optimizing users' energies we obtain the code rate gain in comparison with time division.

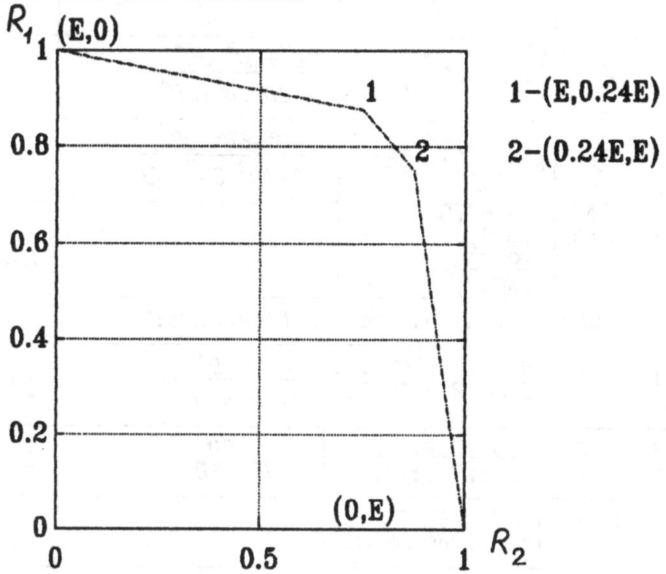

Fig.3 Region of the achievable codes rates ($P_{Bk} \leq 10^{-4}$, $H \leq 2^7$, $q = 2$, $E/N_0 = 13$ dB)

In [8] we considered two classes of low rate q-ary ($q>2$) convolutional codes. The classes of the $(m,m+1,1/m,2^m)$ and $(m+1,m+2,1/m,2^m)$, $m=2,3,\ldots$ codes are presented there. These codes are defined by the matrix $G^{**} = \| G_0^T \ldots G_\nu^T \|$ of size $m\times(\nu+1)$, where T means transpose and G_l, $\overline{l=0,\nu}$, denotes the binary matrices of size $1\times m$ determined by the connections of the l-th delay element with adders by modulo-2 .

For the $(m,m+1,1/m,2^m)$ codes G^{**} is given in the form $G^{**} = \| 1_m | I_m \|$, where 1_m denotes the binary column with one on the m-th position only. For the $(m+1,m+2,1/m,2^m)$ codes matrix G^{**} is equal to

$G^{**}=\| e|1_m|I_m\|$, where e denotes the binary column of all ones. Expressions for transfer functions and their derivatives for these two classes of codes are presented in [8]. The described codes have the best spectrum among codes with the same parameters.

The detailed tables for high rate (2/3, 3/4) q-ary ($q>2$) convolutional codes and their performances, obtained by exhaustive computer search, are given in [9]. Below we present a reduced table for such codes. Here as G_i, $i=\overline{0,\nu}$ we denote binary matrices of size

Table

Codes with code rate 3/2 bits/signal

q,ν,d_f	$G_0 \ldots G_\nu$	T_i, F_i, $i=\overline{d_f,d_{f+4}}$	$T(D)$	H
4,1,2	6 0 2 8 1 4	7,16,79,270,... 11,52,319,1480,...	$\frac{7D^2+2D^3-2D^4}{1-2D-7D^2+2D^4}$	2^4
8,1,2	20 0 2 32 1 8	4,6,25,54,... 4,9,50,141,552,...	$\frac{4D^2+6D^3-3D^4}{1-7D^2-3D^3+3D^4}$	2^4
4,1,3	10 4 3 10 1 5	13,57,334,1833,... 29,256,2078,14899,...		3×2^4

Codes with code rate 2 bits/signal

q,ν,d_f	$G_0 \ldots G_\nu$	T_i, F_i	$T(D)$	H
8,1,2	4 2 2 1	4,15,56,209,... 5,38,214,1068,...	$\frac{4D^2-D^3}{1-4D+D^2}$	2^4
16,1,2	2 8 1 4	3,9,27,81,... 3,18,81,324,...	$\frac{3D^2}{1-3D}$	2^4
8,2,3	2 4 2 6 1 1	5,30,78,1088,.. 7,97,840,6753,...		2^6
16,2,3	10 4 2 6 9 1	3,12,54,240,... 3,27,177,1044,...	$\frac{3D^3}{1-4D-2D^3}$	2^6
16,3,4	8 2 12 9 4 1 6 8	3,15,70,... 3,34,240,...		2^8
32,2,3	1 4 16 2 8 1	3,10,41,163,.. 3,20,115,598,...	$\frac{3D^3+D^4-D^5}{1-3D-4D^2+D^4}$	2^6

Codes with code rate 3 bits/signal

q,ν,d_f	$G_0 \ldots G_\nu$	T_i, F_i	$T(D)$	H
16,1,2	8 4 4 2 2 1	11,116,1222,.. 16,340,5387,...	$\frac{11D^2-5D^3+D^4}{1-11D+5D^2-D^3}$	2^6
16,2,3	5 8 8 15 2 4 4 1 2	16,231,... 28,866,...		2^9

$(j \times nm)$, where j is the number of input bits and n is the number of output 2^m-ary signals.

Matrices G_i are determined by the connections of the i-th delay elements belonging to j binary registers with adders by modulo-2. In the table we give the decimal equivalents for the rows of matrices G_i; T_i, F_i, $i=\overline{d_f, d_{f+4}}$ are weight spectrum coefficients corresponding to the transfer function $T(D)$ and its derivative $F(D)$ respectively. The decoding complexity H is measured as the number of nodes in trellis diagram multiplied by the number of comparisons one should fulfill in each node.

Using codes from [9] one can obtain bounds on achievable code rates analogous to those ones for binary codes.

References

[1] P.Mathys. "A class of codes for a T active users out of N multiple-access communication system", IEEE, Trans. Inform. Theory, vol.36, no.6, Nov.1990.

[2] R.Lupas and S.Verdu. "Linear multiuser detectors for synchronous code-division multiple-access channels", IEEE, Trans. Inform. Theory, vol.35, no.1, pp.123-136, Jan.1989.

[3] S.Verdu. "Minimum probability of error for asynchronous Gaussian multiple-access channels", IEEE Trans. Inform. Theory, vol.IT-32, no.1, pp.85-96, Jan.1986.

[4] S.C.Chang, E.J.Weldon,Jr. "Coding for T-user multiple-access channels", IEEE Trans. Inform. Theory, vol.IT-25, no.6, p.684-691, Nov.1979.

[5] S.C.Chang. "Further results on coding for T-user multiple-access channels", IEEE Trans. Inform. Theory, vol. IT-30, no.2, pp.411-415, Mar.1984.

[6] I.E.Bocharova, F.A.Taubin."Generalized correlation metric for decoding in a multiple-access channel", Problems Info. Trans., vol.26, no.2, pp.38-44, 1990.

[7] I.E.Bocharova. "Generating functions for a class of high rate convolutional codes," II International Workshop "Algebraic and combinatorial coding theory," 16-22 September,1990, pp.38-41.

[8] I.E.Bocharova. "Almost orthogonal codes for orthogonal signals," Fifth joint Soviet-Swedish International Workshop on Information theory, "Convolutional codes; multi-user communication," 13-19 January, 1991, pp.25-28.

[9] I.E.Bocharova, B.D.Kudryashov "Convolutional codes for data transmission using orthogonal signals," Proc. International Symposium "Satellite communication: Present and Future," October 1-5, 1990, pp.b.12.1-b.12.10.

[10] A.N.Trofimov, F.A.Taubin "Bound on decoding error probability for channel with multiple access and interference," Problems Info. Trans., vol.22, no.3, pp.3-15, 1986.

[11] G.C. Clark,Jr. and J. Cain "Error-correction coding for digital communications," Plenum Press, New York, 1981.

[12] A.J.Viterbi, J.K.Omura "Principles of digital communications and coding," New York McGraw-Hill Book company,1979.

[13] A. J. Viterbi, I. M. Jacobs "Advances in coding and modulation for non-coherent channels affected by fading, partial band and multiple-access interference," Advances in communication systems, vol.4, New York: Academic, pp.279-308, 1975.

Lecture Notes in Computer Science

For information about Vols. 1–481
please contact your bookseller or Springer-Verlag

Vol. 524: G. Rozenberg (Ed.), Advances in Petri Nets 1991. VIII, 572 pages. 1991.

Vol. 525: O. Günther, H.-J. Schek (Eds.), Advances in Spatial Databases. Proceedings, 1991. XI, 471 pages. 1991.

Vol. 526: T. Ito, A. R. Meyer (Eds.), Theoretical Aspects of Computer Software. Proceedings, 1991. X, 772 pages. 1991.

Vol. 527: J.C.M. Baeten, J. F. Groote (Eds.), CONCUR '91. Proceedings, 1991. VIII, 541 pages. 1991.

Vol. 528: J. Maluszynski, M. Wirsing (Eds.), Programming Language Implementation and Logic Programming. Proceedings, 1991. XI, 433 pages. 1991.

Vol. 529: L. Budach (Ed.), Fundamentals of Computation Theory. Proceedings, 1991. XII, 426 pages. 1991.

Vol. 530: D. H. Pitt, P.-L. Curien, S. Abramsky, A. M. Pitts, A. Poigné, D. E. Rydeheard (Eds.), Category Theory and Computer Science. Proceedings, 1991. VII, 301 pages. 1991.

Vol. 531: E. M. Clarke, R. P. Kurshan (Eds.), Computer-Aided Verification. Proceedings, 1990. XIII, 372 pages. 1991.

Vol. 532: H. Ehrig, H.-J. Kreowski, G. Rozenberg (Eds.), Graph Grammars and Their Application to Computer Science. Proceedings, 1990. X, 703 pages. 1991.

Vol. 533: E. Börger, H. Kleine Büning, M. M. Richter, W. Schönfeld (Eds.), Computer Science Logic. Proceedings, 1990. VIII, 399 pages. 1991.

Vol. 534: H. Ehrig, K. P. Jantke, F. Orejas, H. Reichel (Eds.), Recent Trends in Data Type Specification. Proceedings, 1990. VIII, 379 pages. 1991.

Vol. 535: P. Jorrand, J. Kelemen (Eds.), Fundamentals of Artificial Intelligence Research. Proceedings, 1991. VIII, 255 pages. 1991. (Subseries LNAI).

Vol. 536: J. E. Tomayko, Software Engineering Education. Proceedings, 1991. VIII, 296 pages. 1991.

Vol. 537: A. J. Menezes, S. A. Vanstone (Eds.), Advances in Cryptology – CRYPTO '90. Proceedings. XIII, 644 pages. 1991.

Vol. 538: M. Kojima, N. Megiddo, T. Noma, A. Yoshise, A Unified Approach to Interior Point Algorithms for Linear Complementarity Problems. VIII, 108 pages. 1991.

Vol. 539: H. F. Mattson, T. Mora, T. R. N. Rao (Eds.), Applied Algebra, Algebraic Algorithms and Error-Correcting Codes. Proceedings, 1991. XI, 489 pages. 1991.

Vol. 540: A. Prieto (Ed.), Artificial Neural Networks. Proceedings, 1991. XIII, 476 pages. 1991.

Vol. 541: P. Barahona, L. Moniz Pereira, A. Porto (Eds.), EPIA '91. Proceedings, 1991. VIII, 292 pages. 1991. (Subseries LNAI).

Vol. 543: J. Dix, K. P. Jantke, P. H. Schmitt (Eds.), Nonmonotonic and Inductive Logic. Proceedings, 1990. X, 243 pages. 1991. (Subseries LNAI).

Vol. 544: M. Broy, M. Wirsing (Eds.), Methods of Programming. XII, 268 pages. 1991.

Vol. 545: H. Alblas, B. Melichar (Eds.), Attribute Grammars, Applications and Systems. Proceedings, 1991. IX, 513 pages. 1991.

Vol. 547: D. W. Davies (Ed.), Advances in Cryptology – EUROCRYPT '91. Proceedings, 1991. XII, 556 pages. 1991.

Vol. 548: R. Kruse, P. Siegel (Eds.), Symbolic and Quantitative Approaches to Uncertainty. Proceedings, 1991. XI, 362 pages. 1991.

Vol. 550: A. van Lamsweerde, A. Fugetta (Eds.), ESEC '91. Proceedings, 1991. XII, 515 pages. 1991.

Vol. 551:S. Prehn, W. J. Toetenel (Eds.), VDM '91. Formal Software Development Methods. Volume 1. Proceedings, 1991. XIII, 699 pages. 1991.

Vol. 552: S. Prehn, W. J. Toetenel (Eds.), VDM '91. Formal Software Development Methods. Volume 2. Proceedings, 1991. XIV, 430 pages. 1991.

Vol. 553: H. Bieri, H. Noltemeier (Eds.), Computational Geometry - Methods, Algorithms and Applications '91. Proceedings, 1991. VIII, 320 pages. 1991.

Vol. 554: G. Grahne, The Problem of Incomplete Information in Relational Databases. VIII, 156 pages. 1991.

Vol. 555: H. Maurer (Ed.), New Results and New Trends in Computer Science. Proceedings, 1991. VIII, 403 pages. 1991.

Vol. 556: J.-M. Jacquet, Conclog: A Methodological Approach to Concurrent Logic Programming. XII, 781 pages. 1991.

Vol. 557: W. L. Hsu, R. C. T. Lee (Eds.), ISA '91 Algorithms. Proceedings, 1991. X, 396 pages. 1991.

Vol. 558: J. Hooman, Specification and Compositional Verification of Real-Time Systems. VIII, 235 pages. 1991.

Vol. 559: G. Butler, Fundamental Algorithms for Permutation Groups. XII, 238 pages. 1991.

Vol. 560: S. Biswas, K. V. Nori (Eds.), Foundations of Software Technology and Theoretical Computer Science. Proceedings, 1991. X, 420 pages. 1991.

Vol. 561: C. Ding, G. Xiao, W. Shan, The Stability Theory of Stream Ciphers. IX, 187 pages. 1991.

Vol. 562: R. Breu, Algebraic Specification Techniques in Object Oriented Programming Environments. XI, 228 pages. 1991.

Vol. 563: A. Karshmer, J. Nehmer (Eds.), Operating Systems of the 90s and Beyond. Proceedings, 1991. X, 285 pages. 1991.

Vol. 564: I. Herman, The Use of Projective Geometry in Computer Graphics. VIII, 146 pages. 1992.

Vol. 565: J. D. Becker, I. Eisele, F. W. Mündemann (Eds.), Parallelism, Learning, Evolution. Proceedings, 1989. VIII, 525 pages. 1991. (Subseries LNAI).

Vol. 566: C. Delobel, M. Kifer, Y. Masunaga (Eds.), Deductive and Object-Oriented Databases. Proceedings, 1991. XV, 581 pages. 1991.

Vol. 567: H. Boley, M. M. Richter (Eds.), Processing Declarative Kowledge. Proceedings, 1991. XII, 427 pages. 1991. (Subseries LNAI).

Vol. 568: H.-J. Bürckert, A Resolution Principle for a Logic with Restricted Quantifiers. X, 116 pages. 1991. (Subseries LNAI).

Vol. 569: A. Beaumont, G. Gupta (Eds.), Parallel Execution of Logic Programs. Proceedings, 1991. VII, 195 pages. 1991.

Vol. 570: G. Schmidt, R. Berghammer (Eds.), Graph-Theoretic Concepts in Computer Science. Proceedings, 1991. VIII, 253 pages. 1992.

Vol. 571: J. Vytopil (Ed.), Formal Techniques in Real-Time and Fault-Tolerant Systems. Proceedings, 1992. IX, 620 pages. 1991.

Vol. 572: K. U. Schulz (Ed.), Word Equations and Related Topics. Proceedings, 1990. VII, 256 pages. 1992.

Vol. 573: G. Cohen, S. N. Litsyn, A. Lobstein, G. Zémor (Eds.), Algebraic Coding. Proceedings, 1991. X, 158 pages. 1992.